Wastewater treatment and technology

C F Forster

Thomas Telford

Published by Thomas Telford Publishing, Thomas Telford Ltd, 1 Heron Quay, London E14 4JD.
URL: http://www.thomastelford.com

Distributors for Thomas Telford books are
USA: ASCE Press, 1801 Alexander Bell Drive, Reston, VA 20191-4400, USA
Japan: Maruzen Co. Ltd, Book Department, 3–10 Nihonbashi 2-chome, Chuo-ku, Tokyo 103
Australia: DA Books and Journals, 648 Whitehorse Road, Mitcham 3132, Victoria

First published 2003

Also available from Thomas Telford Books
Design of water resources systems. P. Purcell. ISBN 0 7277 3098 3
Basic water treatment 3rd edition. C. Binnie, M. Kimber and G. Smethurst. ISBN 0 7277 3032 0

A catalogue record for this book is available from the British Library

ISBN: 0 7277 3229 3
© Thomas Telford Limited 2003

All rights, including translation, reserved. Except as permitted by the Copyright, Designs and Patents Act 1988, no part of this publication may be reproduced, stored in a retrieval system or transmitted in any form or by any means, electronic, mechanical, photocopying or otherwise, without the prior written permission of the Publishing Director, Thomas Telford Publishing, Thomas Telford Ltd, 1 Heron Quay, London E14 4JD.

This book is published on the understanding that the author is solely responsible for the statements made and opinions expressed in it and that its publication does not necessarily imply that such statements and/or opinions are or reflect the views or opinions of the publishers. While every effort has been made to ensure that the statements made and the opinions expressed in this publication provide a safe and accurate guide, no liability or responsibility can be accepted in this respect by the author or publishers.

Typeset by Academic & Technical, Bristol
Printed and bound in Great Britain by MPG Books, Bodmin

Books are to be returned on or before
the last date below.

7 – DAY LOAN

LIBREX–

WITHDRAWN

LIVERPOOL JMU LIBRARY

3 1111 01305 1667

Contents

Chapter 1. Introduction   1
Pollution, 1
Controlling pollution, 3
    Consents, 3
    Urban pollution management, 10
    Treatment, 14
Basic biology, 14
    Bacteria, 16
    Protozoa, 21
    Viruses, 23
References, 23

Chapter 2. Pre-treatment   25
Preliminary processes, 25
    Screens, 25
    Grit removal, 29
    Storm water, 32
Primary settlement, 34
Hydro-dynamic separators, 41
Sea outfalls, 41
References, 43

Chapter 3. Trickling filters   44
Introduction, 44
Biofilm, 45
Media, 47

Wastewater treatment and technology

Ventilation, 48
Distributors, 49
Operation of trickling filters, 50
    Modes of operation, 50
    Settlement tanks, 55
    Performance, 55
Operational problems, 56
    Ponding, 56
    Flies, 57
    Spring sloughing, 58
    Temperature, 58
    High-rate filters, 60
Plastic media, 60
    High-rate filtration, 60
    Low-rate filtration, 63
Models, 64
References, 65

Chapter 4. Activated sludge process   67

Introduction, 67
Process variables, 70
    Organic loading rate, 70
    Sludge age, 71
    Hydraulic retention time, 72
Aeration plant design, 73
    Aeration devices, 73
    Aerator efficiencies, 77
    Oxygen transferred by aerators, 79
    Oxygen requirements, 79
    Optimization of aeration efficiency, 80
    Dissolved oxygen control, 82
    Hybrid aeration, 83
Sludge settleability, 85
Design of final settlement tanks, 87
Aeration tank configuration, 90
Sludge production, 93
    Sludge wastage rate, 94
    Sludge recycle rate, 94
Operational problems, 95
    Filamentous species, 95
    Bulking, 99

Foam formation, 103
Mathematical models, 105
Sequence for design, 107
References, 108

Chapter 5. Bio-oxidation — other aspects — 113
Introduction, 113
Biological contactors, 113
Submerged filters, 117
    Biological aerated filters, 118
    Submerged aerated filters, 122
Deep shaft process, 122
Sequencing batch reactors, 126
Use of pure oxygen, 128
    Open tanks, 130
    Covered aeration tanks, 133
    Oxygen supply, 135
Oxidation ditches, 137
Reed beds, 142
Membrane reactors, 145
Odours and odour control, 148
    General overview, 148
    Odour measurement, 149
    Dealing with odours, 150
References, 154

Chapter 6. Nutrient removal — 160
Introduction, 160
Nitrogen removal, 161
    Nitrification, 161
    De-nitrification, 164
Phosphorus removal, 168
    Chemical phosphorus removal, 168
    Biological phosphorus removal, 173
Phosphorus recovery, 178
References, 180

Chapter 7. Tertiary treatment — 183
Introduction, 183
Solids' removal, 183
    Lagoons, 183

Grass plots, 184
Upflow clarifiers, 186
Microstrainers, 187
Rapid gravity filters, 188
Deep bed filters, 192
Upflow sand filters: the Dynasand filter, 193
Reed beds, 194
Disinfection, 195
Chlorine, 196
Ozone, 198
Ultraviolet radiation, 200
References, 202

Chapter 8. Sludge handling and disposal — 204
Introduction, 204
Sludge thickening, 209
Gravity thickening, 209
Centrifugation, 210
Belt thickeners, 211
Dissolved air flotation, 212
Sludge rheology, 213
Anaerobic digestion, 216
Essential elements, 216
Design and operation, 218
Dual digestion, 223
Enhancing anaerobic digestion, 225
Thermophilic digestion, 227
Enzymatic hydrolysis, 227
Enhanced treatment options, 228
Aerobic digestion, 228
Lime treatment, 230
De-watering, 230
Filter press, 231
Belt press, 236
Sludge disposal, 237
Disposal to agricultural land, 237
Incineration, 241
Landfill, 245
Composting, 245
Thermal drying, 247
References, 249

Chapter 9. Anaerobic digestion of wastewaters   253
Introduction, 253
Microbiology, 253
Reactor design, 255
    Contact digester, 255
    Anaerobic filter, 257
    Upflow sludge blanket reactor, 258
    Expanded and fluidized bed digesters, 263
    Anaerobic baffled reactor, 264
To digest or not to digest, 265
    Costs, 265
    Comparison with aerobic processes, 265
    Basic considerations, 266
    Modes of operation, 268
Prevalence of anaerobic digestion, 269
References, 270

Chapter 10. Industrial wastewater treatment   273
Introduction, 273
Integrated pollution control, 274
Industrial wastewaters, 275
    Biological treatment, 275
    Enhanced biological treatment, 282
    Physico-chemical treatment, 284
    Enhanced oxidation, 289
    Recovery, 291
Agricultural industry, 294
References, 297

Chapter 11. The future   302
Trace organics, 302
    Endochrine disrupting chemicals, 302
    Toxicity, 304
Sewage biosolids, 305
Activated sludge process, 312
References, 315

Index   319

Preface

There is a tendency to view wastewater treatment as a relatively modern technology. However, there is evidence that systems for the handling of human wastes existed 5000 or more years ago. Perhaps the most notable are at the city of Mohenjo-Dara in Pakistan and the palace of Minos on the island of Crete. Subsequent systems of disposal relied on streets, rivers, cesspits and even the castle moat. As well as creating public health problems there were other potential problems with this approach. Cesspits were frequently built underneath houses and it was not unknown for people to be drowned when floors collapsed.

It took a series of cholera epidemics, the initial development of the science of microbiology and the recognition of how water-borne diseases could be transmitted to start a wastewater treatment technology which was the forerunner of the industry we know at the beginning of the twenty-first century. Indeed, it could be argued that the processes which were developed in the late nineteenth and early twentieth century, namely the trickling filter and the activated sludge process, are still the mainstay of our technologies for wastewater treatment. To be more precise, processes based either on biofilms or on a flocculated biomass are the mainstays of our technologies for wastewater treatment.

There are now a variety of biofilm-based options for treating wastewater. Plastic, in the form of high voidage structures, granules or discs, is now available as a support material for biofilms, and processes may be operated in an aerobic or an anaerobic mode. Similarly, floc-based reactors may now be configured around deep shafts or may use membranes to separate the treated wastewater. Again, some of these may be operated as either aerobic or anaerobic reactors. More is now being required in terms of what a wastewater treatment system has to achieve. Ammoniacal-nitrogen has to be oxidized, nitrates may have to be removed and the concentration of phosphates in the final effluent may be restricted. In addition, the feed to the plants is likely to contain industrial wastewater which may contain an increasingly complex cocktail of chemicals, some of which may be refractory, leading to high residual COD concentrations, some of which may be toxic.

The processing and disposal of sewage sludge has, arguably, been subjected to the least degree of change, at least over the last forty years to the time of writing. However, attitudes are changing. The introduction of the Safe Sludge Matrix coupled with greater environmental awareness has created an increased interest in both new options for processing the sludge and new disposal routes.

Any engineer or technologist wishing to enter the 'Wonderful World of Wastewater Treatment' must be aware of the options available for treatment and must understand how they work, what they can achieve, how they can malfunction and what might be done when they do. This necessitates an appreciation, if not a full understanding, of chemistry, biochemistry and biology, including genetics, as well as engineering and environmental awareness. This book aims to provide some of the information necessary to achieve this.

Acknowledgements

I would like to acknowledge the considerable amount of help (information, comment and criticism) I have received from the UK Water Industry and my colleagues. In particular, Dr R. W. Crabtree (Water Research Centre), Dr C. M. Carliell-Marquet and Mr H. A. Hawkes (University of Birmingham), and Mr I. Gray (Severn Trent Water). Also my wife, Romana, for, once again, ensuring that my words conformed to an acceptable form of English.

1
Introduction

Pollution

Civilization equates to pollution. This is an extreme statement and one which is only partially true. Nevertheless, as civilization has developed so has industrialization and urbanization, and this has resulted in increased water usage and the use of chemicals of increasing complexity. Aquatic pollution can be considered in a number of ways but in simple terms it is the addition of an array of elements to the water. The most important are carbon, nitrogen, phosphorus and heavy metals. When organic carbon compounds are present in water they have the potential for being degraded by bacteria. This process requires oxygen. The way of judging the extent of any pollution by carbonaceous compounds, therefore, is to measure the oxygen demand of the water. This is done as the biochemical oxygen demand (BOD) or the chemical oxygen demand (COD) (Clesceri et al., 1998). Nitrogen will occur as a pollutant in the form of ammoniacal-nitrogen, nitrate or organic nitrogen. Ammoniacal-nitrogen can also exert an oxygen demand as it is oxidized to nitrate but, as free ammonia, it can be toxic to fish. Nitrate can be a problem if the water is abstracted for conversion into potable water and it will also help in promoting eutrophication in surface waters. Soluble orthophosphate is the form of phosphorus which most commonly occurs as a pollutant and will also promote eutrophication.

As water is used it becomes polluted in two ways, by domestic use and by industrial use. The amount of polluting matter generated by a single human is

$$BOD = 55-60 \text{ g} \qquad SS = 60 \text{ g}$$

Ammoniacal-nitrogen = 8–12 g PO_4 = 2–4 g as P

where SS means suspended solids. The polluting matter is dissolved or suspended in the quantity of water deemed to be used by each person daily. This varies from country to country and, sometimes, within a country. The typical value in the UK is 150–170 litres.

Table 1.1. Quality of industrial wastewaters

Industry	BOD (mg/litre)	COD (mg/litre)	SS (mg/litre)	COD:BOD
Landfill leachate	170	1400	270	8·2
Thermomechanical pulping	500	4700	385	9·4
Potato processing	2000	3500	2500	1·75
Distillery	7000	10 000	low	1·43
Papermill[a]	960	3680	639	3·83
Coke oven liquor	1200	3900	950	3·25
Municipal sewage — unsettled	246	590	345	2·40

[a] After primary settlement

Aquatic pollution by industry is more complex and its nature will depend on the industrial processes involved. As can be seen in Table 1.1, the wastewaters from industry can have quite high BOD values and, thus, are thought of as being highly polluting. Compared with municipal sewage they can also have a high COD:BOD ratio. The value for municipal sewage tends to be between 1·25 and 2·5 (Gray, 1999), whereas, as can be seen from Table 1.1, some industrial wastewaters can have a ratio appreciably greater than 3. This means that there is a significant amount of carbonaceous material in the wastewater which can be oxidized chemically but not biologically and, therefore, will not readily be treated by the processes normally available at municipal sewage treatment works (Chapter 10). Industrial wastewaters may also contain compounds which, although only present at low concentrations, may be toxic or recalcitrant, that is to say not readily degraded.

Some industrial wastewaters will be predominantly inorganic in nature and contain heavy metal salts (Table 1.2). These will need to be treated by processes other than those of a biological nature (Chapter 10).

When mixed domestic and industrial wastewaters are being considered or described, for example, in the design or the extension of a sewage treatment works, they are usually described as population equivalent (pe). This number is derived by dividing the total mean

Table 1.2. Composition of a typical untreated metal finishing effluent (Wild, 1987)

Metal	Concentration (mg/litre)
Copper	3–5
Chromium	5–10
Nickel	5–10
Zinc	3–5
Suspended solids	10–50
Cyanide	1–5

daily load (g/d) being handled (BOD concentration expressed in mg/litre multiplied by the flow expressed in m^3/d) by 55 (the daily per capita mass of BOD).

Controlling pollution

Consents

Although legislation plays its part, the first stage in any engineered pollution control strategy is to ensure that polluted waters are collected in a sewerage system so that they can be transported to a treatment works. In the UK there are some 347 000 km of sewers, which transport a daily volume of 11 billion litres to about 9000 sewage treatment works. The discharge of a treated effluent to a river may only be made if it (the discharge) has a licence, called the consent. This is issued by the Environment Agency and will

- take into account the river quality objectives (RQO) of the stretch of river receiving the discharge;
- ensure that there is compliance with the European Urban Wastewater Treatment (UWWT) Directive and other appropriate Directives, e.g. the Nitrate Directive.

The RQO, in effect, describes the use of that stretch of river, for example, abstraction for drinking water, coarse fishing or cattle watering. Most objectives have specific standards associated with the water use. Frequently, these are specified by EU Directives. Table 1.3, for example, shows some of the requirements for waters being used for potable water abstraction.

Table 1.3. *Some of the standards specified for potable water abstraction by the Surface Water Abstraction Directive (75/440/EEC)*

	A1		A2		A3	
	G	I	G	I	G	I
BOD (mg/litre)	–	<3	–	<5	–	<7
Nitrate (mg/litre)	25	50	–	50	–	50
Ammonia (mg/litre)	0·05		1	1·5	2	4
Phosphate (mg/litre)	0·4		0·7		0·7	
Phenol (mg/litre)	–	0·001	0·001	0·005	0·01	0·1
Pesticides (mg/litre)	–	0·001	–	0·002 5	–	0·005
Faecal coliforms/100 ml	20		2000	–	20 000	–
Salmonella		nil/5 litre	nil/litre			

The values of the consent parameters can be derived from a mass balance of the flows and concentrations involved.

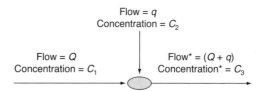

Flow* and concentration* are at the point of mixing and the mass balance at this point gives

$$(Q \times C_1) + (q \times C_2) = [(Q + q) \times C_3)] \tag{1.1}$$

This enables C_2 to be calculated. It is a simplistic approach and computer models can also be used.

Normally, a consent will be based on a 95 percentile compliance and will specify a maximum (absolute) value for each of the parameters required. The parameters specified in a typical consent are suspended solids, BOD and ammoniacal-nitrogen. In some cases the concentrations of total nitrogen and total phosphorus are also specified. In certain cases, if a prescribed substance were present in the raw wastewater, from an industrial discharge to a sewer, a more restrictive consent might be applied.

The UWWT Directive (91/271/EEC) specifies

- that a sewerage system be provided;
- the general standard of treatment;

Introduction

- the minimum sampling frequency and compliance;
- the effluent quality.

There are different criteria for the different types of controlled water (river, estuarine, sea) and for the different sizes of works (Table 1.4). The UWWT Directive, for the first time in the UK, specifies a control level for COD, and also specifies control concentrations for N and P for discharges into sensitive waters (Table 1.5). The Environment Agency specifies which waters are sensitive (Chapter 6).

Table 1.4. Treatment required by UWWT Directive

Population equivalent	Inland	Coastal	Estuarine
<2000	AP	AP	AP
2000–10 000	ST by 2005	ST by 2005[a]	AP
10 000–15 000	ST by 2005	ST by 2005	ST by 2005[a]
15 000–150 000	ST by 2000	ST by 2000	ST by 2000[a]
>150 000	ST by 2000	ST by 2000	ST by 2000

AP is appropriate treatment to meet quality objectives, ST is secondary treatment
[a] Primary treatment in less sensitive areas: SS removal = 50%; BOD removal = 20%

There are other European Directives which may affect the criteria specified in the consent. These include

- Dangerous Substances Directive (76/464/EEC);
- Nitrate Directive (91/676/EEC);
- Freshwater Fish Directive (78/656/EEC);
- Bathing Water Directive (76/160/EEC);
- Surface Water Directive (75/440/EEC).

Table 1.5. Standards required by UWWT Directive

Parameter	Concentration (mg/litre)	Reduction (%)
BOD	25	70–90
COD	125	75
SS[a]	35	90
Total P	2 (<100 000)	80
	1 (>100 000)	
Total N	15 (<100 000)	70–80
	10 (>100 000)	

[a] Optional

Table 1.6. Heavy metals in List I and List II

List I	List II
Cadmium	Chromium
Mercury	Copper
	Lead
	Nickel
	Zinc

For example, in accordance with the Dangerous Substances Directive, there are statutory environmental quality standards (EQSs) in the UK for List I substances (Table 1.6). These are based on annual means (Table 1.7). There are also operational environmental quality standards for List II substances, which include a range of heavy metals. These, together with mercury and cadmium which are included in List I, are all potentially harmful to the environment, although it must be remembered that environmental conditions — temperature, pH, water hardness — will affect their toxicities. The general effects of some of the heavy metals are listed in Tables 1.8 and 1.9.

Discharges from industry to sewer must be licensed (consented). Although, in principle, trade effluent control has changed little since 1936/7, the main legislation in 2003 is the Water Industry Act, 1991, and the Environmental Protection Act, 1995, together with the Environmental Protection (Control of Dangerous Substances) Regulations and

Table 1.7. Examples of the environmental quality standards required in the UK for List I substances

Substance	Statutory EQS (μg/litre)
Mercury and its compounds	1
Cadmium and its compounds	5
Carbon tetrachloride	12
Pentachlorophenol	2
Aldrin	0·01
Dieldrin	0·01
Endrin	0·005
Trichloroethylene	10
Tetrachloroethylene	10

Table 1.8. *Typical toxicities of heavy metals in fresh water. LC 50 means that the given concentration will be lethal to 50% of the species in the specified time*

Metal	Affected species	Effect	Concentration (mg/litre)
Cadmium	Common carp	96 h LC 50	22
Chromium	Brook trout	96 h LC 50	59
Lead	Brook trout	96 h LC 50	3·4
Nickel	Rainbow trout	96 h LC 50	45
Zinc	Rainbow trout	48 h LC 50	35

the Environmental Protection (Prescribed Processes and Substances) Regulations. This legislation

- introduces the concept of Integrated Pollution Control (Chapter 10);
- defines prescribed processes and special category wastes;
- empowers the Secretary of State to make regulations for the treatment, keeping or disposal of particularly difficult or dangerous wastes.

The Water Industry Act also specifies that for industrial discharges

- discharges to sewer are illegal without a water service company consent;

Table 1.9. *A generalized view of the toxicities of the List I and List II heavy metals*

Metal	Effect
Cadmium	In humans can damage the kidney and bones and is associated with itai–itai disease.
Chromium	Cr (VI) is more toxic than Cr (III). Can be immobilized in the body and accumulate as Cr (III).
Copper	An essential element and the body can regulate its level homeostatically. Large, acute doses can have harmful, even fatal, effects. Is also phytotoxic.
Lead	A cumulative poison which, in humans, can cause hypertension and brain damage.
Mercury	Can exist as an inorganic, divalent form or as methyl mercury. This latter form is the one most commonly digested by humans.
Nickel	Main concern is its phytotoxicity.
Zinc	Present in most natural waters. The main concern is its phytotoxicity.

Table 1.10. *Compounds likely to be restricted in trade effluent consents*

Type of protection	Typical compounds
Sewer personnel	Cyanides, sulphides
Sewer fabric	Petroleum spirit, carbides, sulphates
STW personnel	Cyanides, sulphides
STW operation	List I compounds, heavy metals

- the water service company may control discharges;
- the water service company may make charges;
- special category effluents (which contain prescribed substances or come from prescribed processes) must be referred to the Secretary of State;
- there is a prohibition (Section III) on the discharge of petroleum spirit, calcium carbide and carbon disulphide.

Trade effluent consents are aimed at protecting personnel working in a sewer and the sewer fabric. In addition, they aim to protect personnel at a treatment works and the operation of sewage treatment (Table 1.10).

Consents will, therefore, typically specify

- pH and temperature;
- maximum daily flow and flow-rate;
- concentrations of controlled substances (heavy metals, sulphates, cyanides, solids);
- monitoring requirements;
- charging policy.

Charging is aimed at recovering treatment costs and is usually based on the Mogden formula. Thus, the total cost (pence per cubic metre), C, is given by

$$C = R + P + B\left(\frac{O_T}{O_S}\right) + S\left(\frac{S_T}{S_S}\right) \quad (1.2)$$

where R = reception and conveyance cost, P = preliminary and primary treatment cost, B = bio-oxidation cost, S = sludge disposal costs, O_T = trade effluent COD (after settlement), O_S = sewage COD (after settlement), S_T = trade effluent suspended solids, S_S = crude sewage suspended solids.

The values for these parameters vary from year to year and from company to company (Figs 1.1, 1.2 and Table 1.11). Some water service companies

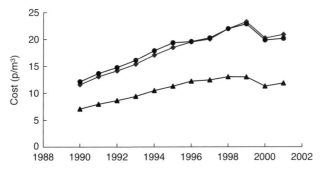

Fig. 1.1. Elements of costing used by one water service company in the UK for B (●), P (◆) and S (▲) in the Mogden formula

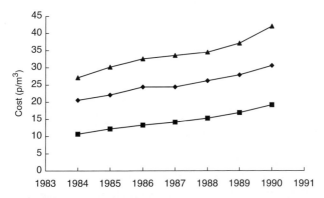

Fig. 1.2. Trade effluent costs for different water service companies

use a modified Mogden formula which includes a factor (M) for the use of a long sea outfall. The way the formula is applied in calculating the cost to industry depends on the processes being used at the sewage treatment works receiving the wastewater. Thus, if treatment involved the use of

Table 1.11. *Examples of the factors used in the Mogden formula (2001/2002)*

	Company 1	Company 2	Company 3
R (p/m^3)	21·14	25·33	35·4
V (p/m^3)	20·81	21·49	32·62
B (p/m^3)	20·18	26·4	57·43
S (p/m^3)	11·86	16·7	34·30
M (p/m^3)	–	3·24	5·94
COD$_S$ (mg/litre)	891	452	744
SS$_S$ (mg/litre)	319	612	489

primary settlement and a long sea outfall (Chapter 2), the terms R, P, S and M would be used, whereas if primary treatment alone was being used only the terms R, P and S would be applied.

Urban pollution management

The Environment Agency has derived a series of River Ecosystem (RE) classes based on the conditions which will sustain different categories of fish populations. These are

- RE1   Very good quality; suitable for all fish;
- RE2   Good quality; suitable for all fish;
- RE3   Fair quality; suitable for high-class coarse fish;
- RE4   Fair quality; suitable for coarse fish;
- RE5   Poor quality; coarse fish populations likely to be limited.

The values which define these classes are given in Table 1.12. They can be used to limit the polluting inputs to a river. The discharges from a sewage treatment works are regular, in that there will always be flow, and at an easily accessible location. This means that their quality can be monitored easily. In wet weather conditions a number of other discharges need to be considered.

There can be discharges from the storm tanks at the sewage works (Chapter 2). There can be discharges from surface water overflows, and combined sewage overflows may be activated (Fig. 1.3). All of these can have an impact on the river water quality, but the impact will be both intermittent and transient and the actual discharges may not easily be monitored. It is recognized that the impacts caused by these inputs during wet weather adversely affect river water quality. This has led to the development of the concept of urban pollution management (UPM), the details of which have been laid in the Urban Pollution Management Manual, which was published in 1994 and updated in 1998 (FWR, 1994, 1998). Urban pollution management has been defined as the management of wastewater discharges from sewer and sewage treatment systems under wet weather conditions such that the requirements of the receiving water are met in a cost-effective way. In other words, the actual conditions in the river are compared with the Fundamental Intermittent Standards (FISs), which are based on dissolved oxygen and free (un-ionized) ammonia concentrations. Urban pollution management identifies values for these parameters for three types of ecosystem (salmonid fishery, cyprinid fishery and marginal cyprinid fishery) and links a concentration to

Table 1.12. Concentrations defining the River Ecosystem classes

Class percentile	DO (%) 10	BOD (mg/litre) 95	Amm.N (mg/litre) 90	Unionized ammonia (mg/litre) 95	pH 5 lower/95 upper	Hardness (mg/litre)	Dissolved Cu (µg/litre) 95	Total Zn (µg/litre) 90
RE1	80	2·5	0·25	0·021	6–9	≤10 10–50 50–100 >100	5 22 40 112	30 200 300 500
RE2	70	4·0	0·60	0·021	6–9	≤10 10–50 50–100 >100	5 22 40 112	30 200 300 500
RE3	60	6·0	1·3	0·021	6–9	≤10 10–50 50–100 >100	5 22 40 112	300 700 1000 2000
RE4	50	8·0	2·5	—	6–9	≤10 10–50 50–100 >100	5 22 40 112	300 700 1000 2000
RE5	20	15·0	9·0	—	—			

Wastewater treatment and technology

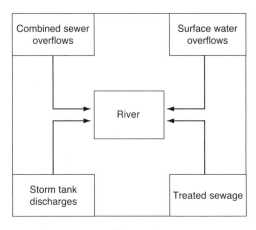

Fig. 1.3. River inputs during wet weather conditions

the duration for which the concentration exists and its return period (Table 1.13) (Clifforde and Crabtree, 2002). For example, in a river supporting cyprinids, the maximum concentration of free ammonia which is permitted for a one-hour period once a month is 0·15 mg/litre, whereas if the return period was a year, the maximum permitted concentration would be 0·25 mg/litre. These limits also depend on the dissolved oxygen concentration, the pH and the temperature, and the UPM manual provides a series of correction factors for adjusting the concentrations. The second edition of the UPM manual allows for 99 percentile values which are related to the RE classes to be considered as an alternative to FIS values. Table 1.14 shows these values for BOD, total ammoniacal-nitrogen and free ammonia.

The UPM procedure can use a series of mathematical models (for example, INFOWORKS, STOAT and MIKE 11) to make predictions about the quality and quantity of flows in the sewers, the sewage works and the river respectively. Alternatively, a simplified modelling tool, SIMPOL, provided within the UPM manual, can be used to assess

Table 1.13. FIS values for a six hour duration period to be exceeded not more than once per year (Clifforde and Crabtree, 2002)

Ecosystem	Dissolved oxygen (mg/litre)	Free ammonia (mg/litre)
Salmonid fishery	4·5	0·04
Cyprinid fishery	4·0	0·15
Marginal cyprinid fishery	2·5	0·20

Table 1.14. 99 percentile values for the different RE classes (Clifforde and Crabtree, 2002)

Class	BOD (mg/litre)	Total ammoniacal-nitrogen (mg/litre)	Free ammonia (mg/litre)
RE1	5·0	0·6	0·04
RE2	9·0	1·5	0·04
RE3	14·0	3·0	0·04
RE4	19·0	6·0	
RE5	30·0	25·0	

the environmental impact on the river of particular options and rainfall conditions. The SIMPOL model, which contains a series of modules to deal with, for example, the sewers and the river subcatchments, is a conceptual model and does not provide the same level of detail as a deterministic model. However, it must be recognized that modelling, at whatever level, does require an appreciable amount of data to enable the models to be calibrated and verified. The UPM procedure

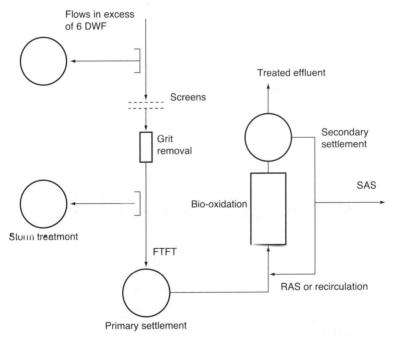

Fig. 1.4. Schematic diagram of a sewage treatment works (RAS and SAS apply to the activated sludge process and recirculation applies to the trickling filter option)

13

Wastewater treatment and technology

Fig. 1.5. Aerial view of a sewage treatment works

was used at Derby, UK, when significant investments in sewerage and sewage treatment were being planned (Crabtree et al., 1996).

Treatment
The layout of a typical sewage treatment works is shown in Figs 1.4 and 1.5 and the main processes are

Process	Operations to be considered
Screens	Solids removal; settlement
Grit removal	Settlement; hydraulics
Storm water separation	Hydraulics
Primary treatment	Settlement
Trickling filters	Bio-oxidation
Activated sludge	Bio-oxidation
Secondary settlement	Settlement

Basic biology
Since the main processes which are used in wastewater treatment are biological, there is a need to consider what microbial species are

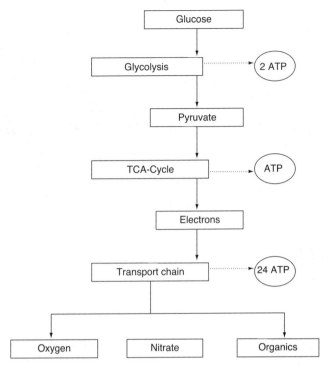

Fig. 1.6. Simplified biochemical pathway for the degradation of glucose

important, and how food is used in cells. Microbial cells use food for energy and growth. The overall process is called metabolism. This can be considered to be two distinct processes: the breakdown of molecules (catabolism) and the re-building of new ones (anabolism). The essential molecules in all these reactions are

- the enzymes, which are the biological catalysts for the cellular reactions. These may be constitutive enzymes, which are always present in the cell, or inducible enzymes, which are synthesized when required;
- energy-rich molecules such as adenosine triphosphate (ATP).

Cells use energy in the form of ATP in the synthesis of new molecules. It is, therefore, important that ATP is regenerated during catabolism. Thus, if the degradation of glucose is considered (Fig. 1.6), it can be seen that ATP can be generated, and that electrons are generated by the tricarboxylic acid (TCA) cycle. These are transferred along an electron transport pathway to one of three final electron acceptors,

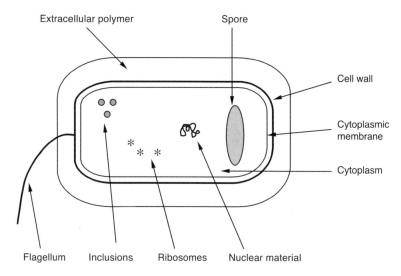

Fig. 1.7. *Schematic diagram of a bacterial cell*

molecular oxygen, inorganically bound oxygen in the form of nitrate or an organic molecule. These relate to aerobic, anoxic and anaerobic conditions respectively.

Bacteria

Bacteria are simple cells which do not have a specific nucleus. Their nuclear material is simply a coiled zone within the cytoplasm (Fig. 1.7). The cytoplasm may also contain a variety of inclusions, such as polyphosphates, poly-β-hydroxybutyric acid or sulphur granules, and ribosomes which are involved in protein synthesis. Bacteria do not contain sub-units dedicated to specific functions analogous to the liver or lungs in mammals. Most of this type of function is carried out by the cytoplastic membrane, which is 7–8 nm thick and has extensive invagination. This increases its surface area. The cell wall is a rigid structure which defines the shape of the cell. Bacteria can occur as rods, spheres (cocci), chains, agglomerates or filaments (Figs 1.8 and 1.9). Those species which have flagella are motile (i.e. they can move). Some species will form a gel-like material outside the cell wall. This is sometimes known as a capsule or a slime layer and is made up of polysaccharide material. This type of material is thought to protect the bacteria from desiccation. It is of interest to the wastewater treatment engineer because it is important to the formation of biofilms

Introduction

Fig. 1.8. Rod-shaped bacteria in activated sludge

(Chapters 3 and 5) and the flocculation of activated sludge (Chapter 4). Some bacteria can also form spores. These are non-growing states which are very resistant to adverse conditions, such as heat, radiation or desiccation, and are little more than nucleic acid depositories. They are produced during the normal cell reproduction cycle and can survive after the main population of vegetative cells has died or been killed. They can germinate when the environment becomes favourable to

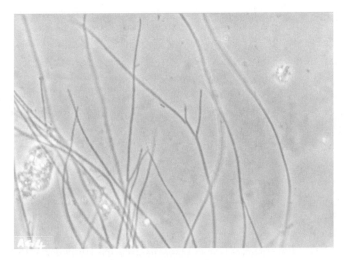

Fig. 1.9. The filamentous bacterium Sphaerotilus natans

Table 1.15. Typical bacterial dimensions

	Diameter (μm)	Length (μm)	Shape
Staphylococcus aureus	0·8–1·0	–	Sphere
Escherichia coli	0·4–0·7	1·0–1·3	Rod
Bacillus anthracis	1·0–1·3	3·0–10·0	Rod

the growth of the vegetative cells. The species which the wastewater treatment engineer needs to know about are clostridia and the anthrax bacillus.

The dimensions of bacteria (length, width, diameter) are only a few micrometres (Table 1.15). They reproduce asexually by doubling. That is to say, each cell will divide to form two. Thus, a batch culture will follow a growth curve similar to that shown in Fig. 1.10. Initially, there is a lag period while the cells adjust to the growth medium and synthesize any inducible enzymes necessary to degrade the substrate being offered. Its length depends on how different the growth medium is from that previously encountered by the cells. This is followed by a logarithmic or exponential phase where the cell numbers increase according to the geometric progression 2^0, 2^1, $2^2, \ldots, 2^n$. Therefore, starting with a culture containing N_O cells, the number of cells after n replications is given by

$$N_n = N_O 2^n \tag{1.3}$$

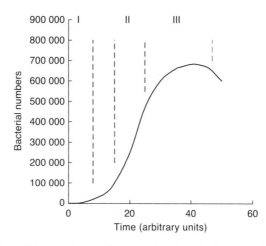

Fig. 1.10. Typical bacterial growth curve showing the lag period (I), the logarithmic phase (II) and the endogenous stage (III)

This can be re-arranged to give

$$n = \frac{\log_{10} N_n - \log_{10} N_O}{\log_{10} 2} \tag{1.4}$$

If the time taken for these n replications is Δt, then the mean generation time (MGT) is given by

$$\text{MGT} = \frac{t}{n} \tag{1.5}$$

The logarithmic phase will continue until one of the nutrients (carbon, nitrogen, phosphorus or oxygen) becomes limiting. When this happens the growth enters the endogenous or stationary phase, where the rate of growth is zero, and ultimately a death phase.

In general, their requirements for growth are

- a carbon source;
- a supply of nitrogen;
- phosphorus;
- trace elements such as K, Mg, Fe, Zn, Cu;
- a pH neutral or at least in the range 5·5–8·5;
- a degree of moisture usually specified as a water availability (solution vapour pressure/vapour pressure of water) $> 0·6$;
- a temperature of about 30 °C for the so-called mesophiles;
- a specific oxygen regime.

The oxygen requirements separate bacteria into those which require oxygen (the strict or obligate aerobes), those which cannot tolerate oxygen (the obligate anaerobes) and those which can grow both with and without oxygen (the facultative anaerobes). There are also bacteria which will tolerate anoxic conditions where there is zero-dissolved oxygen but nitrate is present. In this case, the bacteria use the oxygen bound in the nitrate.

It is now recognized that there can be extremophiles which can grow outside the normal ranges of pH, temperature or salinity. Thus, psycrophilic species can grow at very low temperatures (0–20 °C) and thermophiles grow at high temperatures (40–80 °C).

Bacteria obtain their energy for growth either from carbonaceous material (heterotrophs) or from inorganic compounds (autotrophs). The MGTs vary from species to species and will also be a function of the temperature. From the point of view of wastewater treatment,

Table 1.16. A comparison of the rates of bacterial growth

Species	Temperature (°C)	μ_{max} (d^{-1})	MGT (h)
Escherichia coli	30		0·33
Floc formers[a]	–	5·4	
Zoogloea ramigera	–	5·5	
Sphaerotilus natans	–	0·65	
Actinomycetes[a]	–		4–13
Nitrosomonas	10		82·6
Nitrosomonas	20		31·6
Nitrobacter	10		41·4
Nitrobacter	20		23·1

[a] In mixed liquors

the concern about MGT is the difference between the carbon-oxidizers and the slow growing nitrifiers, and between the floc-formers and the slow growing filamentous bacteria (Table 1.16).

The relationship between growth rate (μ) and substrate concentration (S) is given by the Monod equation

$$\mu = \frac{\mu_{max}(S)}{K_S + (S)} \qquad (1.6)$$

where K_S is the substrate concentration at which μ is half of μ_{max}. This is important for an understanding of how selectors work when discriminating against filamentous bacteria in the activated sludge process (Chapter 4) and for the mathematical modelling of bio-oxidation processes.

In a bio-oxidation plant there are many different species of bacteria and a variety of different substrates. This means that a number of interactions may occur, probably the most important of which is competition. This happens when more than one species is dependent on a specific substrate. Under these circumstances, the faster growing species will outgrow the other species. The variety of substrates, particularly when trade effluents are present, may cause the phenomenon known as diauxie to occur, when one substrate is used preferentially and growth on the second does not take place until the preferred substrate is exhausted (Fig. 1.11). Therefore, if the hydraulic retention time in, for example, the aeration tank of an activated sludge plant is not long enough to encompass the second growth phase, there may be an appreciable residual BOD or COD.

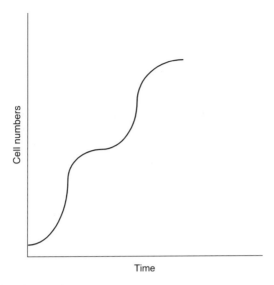

Fig. 1.11. Diauxic growth

Staining is a technique which is used to aid the examination of bacteria. Some stains are specific for individual components (spores, capsules, flagella). Others are a step towards identification. Those which might be used during the operation of a wastewater treatment plant are the Gram stain and the Neisser stain. Both of these are differential stains, deemed to be positive or negative, and are used for the identification of filamentous bacteria (Chapter 4).

Protozoa
Protozoa can be thought of as being microscopic 'animals' which are able to feed on smaller microbes or on soluble organics. The protozoans are a more advanced species than the bacteria and have organs designed for specific functions. Thus, some species have a distinct 'mouth', the cytostome, and the water balance within the cell is controlled by the contractile vacuole. They can be classified on the basis of their feeding habits as

- saprozoic   feed on soluble organics;
- holozoic   feed on bacteria/soluble organics;
- predatory   feed on algae and/or other protozoa.

An alternative viewpoint is that it is probably better to consider their classification on the basis of shape and movement. The amoebae 'glide'

Wastewater treatment and technology

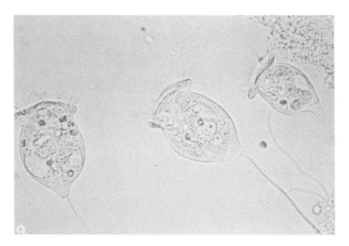

Fig. 1.12. Vorticella, a ciliated protozoa

over surfaces by extending finger-like protrusions, pseudopodia, which are then used as a focus for the rest of the cell to close on to. The free-swimming ciliates, such as *Paramecium*, move by means of their cilia (Fig. 1.12). These are tiny hairs which surround the cell and their movement gives the cell locomotion. The stalked ciliates, such as *Vorticella* (Fig. 1.13), do not have movement since they are attached to surfaces by their stalks. However, their stalks are contractile, which does give them movement within a restricted volume. Flagellates, as the name suggests, move by means of their flagella. Finally, there are the

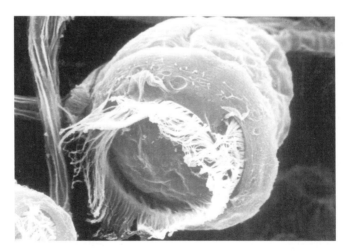

Fig. 1.13. Scanning electron micrograph of a stalked ciliate

Introduction

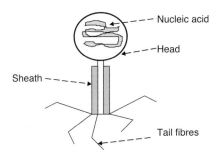

Fig. 1.14. Schematic diagram of a virus

sporozoa which have no organs to give them movement. They can, however, move by flexing their bodies.

From the point of view of management of a wastewater treatment plant, particularly an activated sludge plant, visual classification is probably the better option, as a microscopic evaluation of the sludge will provide an assessment of its quality (Chapter 4).

Viruses

Viruses are very small microbes (the foot and mouth virus, for example, has a diameter of about 10 nm), which are much more resistant to adverse conditions than those species that have been discussed previously. They consist of a nucleic acid core and a protein shell (Fig. 1.14) and are absolute parasites, only being able to reproduce in their host cells. Viruses are host-cell-specific. There are human viruses, animal viruses, plant viruses and even bacterial viruses, the last known as bacteriophages. Bacteriophages have been used as tracers to examine the mixing characteristics of wastewater treatment plants and to follow the movement of effluents in rivers (Sinton and Ching, 1987; Frederick and Lloyd, 1996). It has also been suggested that they could be used to attack unwanted filamentous bacteria in activated sludge plants.

References

Clesceri, L. S., Greenberg, A. E. and Eaton, A. D. (eds) (1998) *Standard Methods for the Examination of Water and Wastewater*, American Public Health Association, American Water Works Association, Water Environment Federation, Washington, DC, USA, 20th edn.

Clifforde, I. and Crabtree, B. (2002) Developments in urban pollution management, *J. CIWEM*, **16**, 229–32.

Crabtree, B., Earp, W. and Whalley, P. (1996) Demonstration of the benefits of integrated wastewater planning for controlling transient pollution, *Wat. Sci. Technol.*, **33(2)**, 209–18.

Foundation for Water Research (1994) *Urban Pollution Management Manual*, FR/CL0009, 1st edn.

Foundation for Water Research (1998) *Urban Pollution Management Manual*, FR/CL0009, 2nd edn.

Frederick, G. L. and Lloyd, B. J. (1996) Evaluation of retention time in waste stabilization ponds using *Serratia marcescens* bacteriophage as tracer, *Wat. Sci. Technol.*, **33(7)**, 49–56.

Gray, N. F. (1999) *Water Technology: An Introduction for Environmental Scientists and Engineers*, Arnold, London.

Sinton, L. and Ching, S. B. (1987) Evaluation of two bacteriophages as sewage tracers, *Wat., Air and Soil Pollut.*, **35**, 347–56.

Wild, J. (1987) Liquid wastes from the metal finishing industry, in *Surveys in Industrial Wastewater Treatment*, Vol. 3, Barnes, D., Forster, C. F. and Hrudey, S. E. (eds), Longman Scientific and Technical, Harlow, 21–64.

2
Pre-treatment

Although the essential part of any wastewater treatment plant is the secondary, biological stage, the processes which precede it do have considerable significance in determining the efficiency of the secondary stage. Some of the pre-treatment options may also be thought of as appropriate treatment, under the terms of the Urban Wastewater Treatment Directive, for coastal or estuarine discharges (Chapter 1).

Preliminary processes

Screens

Screens are the first of the processes at a sewage treatment works. Their objective is to remove gross solids from the flow and there is a wide range of designs for achieving this. Because of this, only the general principles will be discussed. The typical composition of screenings is shown in Table 2.1. Modern screens are cleaned automatically with rakes. The cleaning action is triggered either by timers or by level electrodes. The bar spacing can be quite variable and the amount of screenings which are collected will vary with spacing and the type of catchment (Figs 2.1 and 2.2). As a rule of thumb, there will be $0.01-0.03 \, m^3$/day per 1000 pe. Gross solids are objectionable and their removal from the treatment works will incur a cost. For example, in 1993, the disposal of screenings cost Severn Trent Water over £500 000 (Clay et al., 1996). Therefore, the handling and disposal of screenings does need careful consideration. Some form of washing and de-watering is probably the best route (Newman, 1986) (Figs 2.3

Wastewater treatment and technology

Table 2.1. *Typical composition of screenings (Clay et al., 1996)*

Component	Concentration (% dry weight)
Paper	20–50
Rags	15–30
Plastic	5–20
Rubber	0–5
Vegetable matter	0–5
Faecal material	0–5

and 2.4) and there are a range of systems available commercially which will do this. Some of them effect the de-watering by filtration and pressing, others use centrifugation. In either case, much of the objectionable organic matter is washed out and the de-watering reduces the volume of material which has to be removed from the site (Fig. 2.5). In addition, it should be remembered that removal from the site will entail transport on public roads, which, in turn, means that there should be no free liquid associated with the solids. It has been recommended that the material produced for disposal should

Fig. 2.1. Initial removal of solids on a screen (courtesy of Severn Trent Water)

26

Pre-treatment

Fig. 2.2. Fine (6 mm) screens (courtesy of Anglian Water)

contain at least 25% dry matter (Clay et al., 1996) and, certainly, this can be achieved (Newman, 1986). Gravity sewers tend to produce screenings with a greater amount of faecal contamination than pumped sewers and this may affect the choice of handling equipment (Clay et al., 1996). An alternative solution is to pass the screenings through a macerator to reduce the material into pieces fine enough for them to be returned to the main flow and be removed with the sludge. However, this is not a solution favoured by most plant operators and the presence of these residual pieces can cause problems when sludge is being applied to farmland.

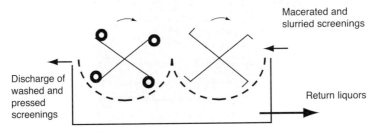

Fig. 2.3. Schematic diagram of a washing/pressing system for screenings

27

Wastewater treatment and technology

Fig. 2.4. Washing/pressing system for screenings

If possible, the screens should be duplicated so that maintenance and the repair of any breakdowns can be carried out without any appreciable disruption in performance. If this cannot be done a by-pass channel should be provided.

Fig. 2.5. Discharge of pressed screenings

The screen chamber will be designed on the basis of

$$W = \left(\frac{X+Y}{Y}\right)\left(\frac{Q}{V \cdot H}\right) \quad (2.1)$$

where W = screen width (m), X = bar width (usually about 15 mm), Y = spacing (mm), Q = maximum flow rate (m³/s), V = maximum velocity (m/s), H = depth of flow at Q (m).

The bar spacing will depend on the use/requirement and a wide range of values are employed but, typically, it would be 75–150 mm at pumping stations, and 15–25 mm for sewage works. This is, essentially, an equation based on experience and the use of non-conventional bar spacings may mean that the relationship is not valid. The type of non-conventional bar spacings which might be encountered would be

- 40–60 mm for primary screens at sewage works;
- 5–13 mm for sea outfalls;
- 12·5 mm at sewage works;
- 3–6 mm for fine screens.

Grit removal

The next stage is grit removal. This is done by reducing the flow velocity to 0·3 m/s so that the inorganic grit settles out in a specific chamber. These chambers can take the form of

- grit channel;
- detritor;
- aerated channels;
- grit traps;
- hydrodynamic separators.

Arguably, the most common approach in the UK is to use constant velocity grit channels (Fig. 2.6) or detritors (Fig. 2.7). For any specific flow, the latter occupy less space and, therefore, would be the preferred option when land was restricted. With constant velocity grit channels, which have a trapezoidal cross-section, the flow velocity is controlled by flumes immediately downstream of the channels. At a velocity of 0·3 m/s, the settlement velocity of grit is about 0·03 m/s. This means that the required length of channel (L) is

$$L = \frac{\text{Depth of flow} \times \text{velocity}}{0 \cdot 03} \quad (2.2)$$

Fig. 2.6. Grit channel with a suction system for the removal of the grit

In other words, a length of 10× the maximum depth of flow is required. However, to accommodate the range of sizes of the grit particles and their different settlement velocities, this value would usually be doubled. The grit which is collected can be removed by the use of dredgers, pumps or suction devices mounted on travelling gantries (Fig. 2.6). Detritors are shallow, square tanks. The flow is directed across the tank by a series of concrete baffles which can be adjusted to ensure that there is an even flow across the full width of the tank (Fig. 2.7). The grit is moved to the periphery of the tank by rotating arms and discharged into a sump for cleaning. Aerated channels, as well as removing grit, will separate any fats, oils and greases (FOG) from the sewage by flotation (Figs 2.8 and 2.9). This can be removed at this stage, an action which can improve the performance of the primary settlement tanks.

Pre-treatment

Fig. 2.7. A detritor

The amount of grit collected will vary with the type of sewerage system, but the rule of thumb is

- for combined systems $4-12 \, m^3$/year per 1000 pe;
- for separate systems   about $4 \, m^3$/year per 1000 pe.

The grit which is collected by any of these techniques is usually disposed of to landfill.

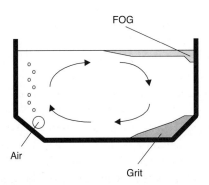

Fig. 2.8. Schematic diagram of an aerated grit channel

Wastewater treatment and technology

Fig. 2.9. Aerated grit channel in operation

Storm water
The final preliminary process is the separation of excess (storm) flow. Some of the excess flow will have been discharged to the river prior to arriving at the works by way of combined sewage overflows, and the maximum flow reaching the works will be restricted by this. However, at some treatment works this is not the case. Even so, there can be, perhaps, an 80-fold variation in the flow arriving at a works. This is too great for hydraulic design and process operation. Therefore, the flow to full treatment is restricted.

Three flows should be identified:

$$\text{dry weather flow (dwf)} = PG + I + E \quad (2.3)$$

$$\text{flow to full treatment (FTFT)} = 3PG + I + 3E \quad (2.4)$$

$$\text{maximum flow arriving at the works} = (PG + I + E) + 1.36P + 3E \quad (2.5)$$

where P = population served (number), G = per capita usage (m^3), I = infiltration (m^3), E = trade effluents (m^3).

Values for these flows are not always easy to obtain. Infiltration can be a particular problem and, if extensions of treatment works are being

Pre-treatment

Fig. 2.10. Empty storm tank

planned, may necessitate measuring the flow at some period of time when there is little or no input from other sources.

Flows in excess of FTFT are separated, usually by a side weir, and taken to storm tanks which are filled sequentially. Storm tanks are usually rectangular horizontal flow tanks (Fig. 2.10), and the first tank in the sequence may be designed without an outlet so that the first foul-flush of storm water, which is notoriously strong, is retained. The normal procedure would be to return settled storm water to the main flow when the input drops to less than FTFT. However, if high flows persist and all the tanks are filled, the flow will be directed into a tank which is already full. This means that settled storm sewage is discharged to the river, which can, in some cases, have the potential for pollution, albeit transitory (Chapter 1). Because of this the discharge may have to comply with a specified quality standard. Therefore, the treatment of storm sewage may be needed. One way of doing this is to use reed beds (Chapter 9) or vortex separators.

The design criteria for storm tanks is not precise. The criteria which have been used are

- 70–78 litres/capita;
- minimum hydraulic retention of 2 hours based on FTFT;

- minimum hydraulic retention of 2 hours based on (maximum flow to works – FTFT).

Primary settlement

Primary tanks use settlement to remove suspended solids (SS) and because of the organic nature of much of the solids, there will also be some removal of BOD. There is no removal of ammoniacal-nitrogen. Typically, the solids removal will be 60–70% and the BOD removal will be 30–35% (Fig. 2.11). Usually, there will be more than one primary tank and it is important to ensure that the flow of sewage is split equally between the tanks. This can be done in several ways, possibly the simplest of which is to divide the channel carrying the sewage progressively (Fig. 2.12) and then to use penstocks to provide fine control.

There are several types of design which can be used for primary tanks, but the two most common are horizontal flow and radial flow.

Horizontal flow tanks will have a length:width ratio of 3:1 to 4:1, a minimum depth of 2 m and a floor slope of 1 in 100. Flow enters and leaves the tank by full width weirs. The final weir should be protected by a scum board. This should be positioned so that the upflow velocity between the board and the weir does not exceed 0·005 m/s. The tanks are fitted with automatic scrapers for sludge removal and there is a sludge hopper at the inlet end of the tank (Figs 2.13 and 2.14). Sludge is withdrawn under the pressure generated by the differential head.

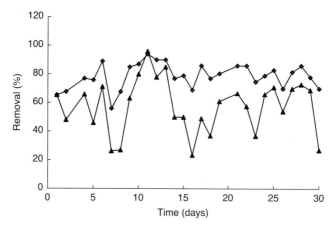

Fig. 2.11. Performance of primary tanks for the removal of BOD (▲) and suspended solids (◆)

Pre-treatment

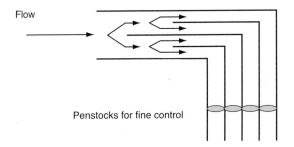

Fig. 2.12. Flow division by channel splitting

Fig. 2.13. Rectangular primary tank with a full-width travelling scraper system

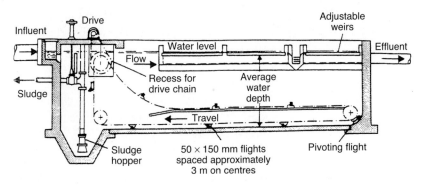

Fig. 2.14. Cross section of a rectangular primary tank

35

Fig. 2.15. *Radial flow tank*

The design is based on maximum flow and the two important criteria are

- hydraulic retention time   2 hours at FTFT;
- surface loading (Q/A)   30–45 m^3/m^2 per day

(Boon and Dolan, 1995),

where A is the plan area of the settlement tanks. Radial flow tanks (Figs 2.15 and 2.16) are circular with diameters of 5–50 m. Their sidewall depth should be 2 m and their floor slope = 7·5–10°. They have a rotating bridge (speed < 1·5 m/min) fitted with scrapers arranged in echelon which move the sludge into a central hopper. The scrapers should be attached to hinged arms so that they can adjust to the profile of the tank floor. Sewage enters the tank through a central bell mouth and stilling chamber and leaves by overflowing a peripheral weir, which should be protected with a scum-board. This should not be located too close to the weir to ensure that the velocities in the vicinity of the weir are not excessive. Ideally, the velocity should not exceed 0·005 m/s.

The design criteria are

- hydraulic retention time   2 hours at FTFT;
- surface loading   45 m^3/m^2 per day

Pre-treatment

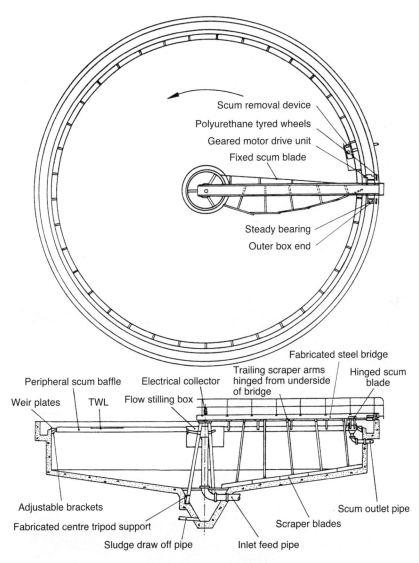

Fig. 2.16. Plan and cross-section of a radial flow primary tank

The design of the stilling chamber also needs consideration. Its role is to reverse the upward velocity of the sewage coming from the inlet bellmouth and re-direct it so that it reaches the floor of the tank where it is deflected radially towards the peripheral weir. In addition, the liquid velocity should be reduced to a value which will not cause any scouring of the settled sludge before it enters the settling zone. Boon and Dolan (1995) have argued that this means that the stilling

37

Wastewater treatment and technology

chamber should have a diameter which is about 10% of the surface area of the tank. This is about 32% of the tank diameter.

Primary settlement will usually generate scum. This is composed of fats, oils and grease mixed with some of the residual solids. This means that the weirs need to be protected by a scum board and that some arrangement must be made to remove the scum from the surface. Scum removal can be achieved by attaching a skimming blade to the travelling bridge. With radial flow tanks, the scum moves to the periphery of the tank where it is removed through a scum trough (Fig. 2.17).

Fig. 2.17. Scum troughs

Fig. 2.18. Use of multiple weirs

With either type of settlement tank, weir loadings (m^3/m per day) should be considered. In general, these should be between 100 and 450 m^3/m per day at maximum flow. If the weir overflow rate is less than 100 m^3/m per day at maximum flow, at lower flows surface tension can affect flow over the weir, while if the weir overflow rate is greater than 450 m^3/m per day, solids can be scoured over the weir. However, primary tanks have operated successfully at appreciably higher rates. Strict adherence to these limits can mean that the diameters of radial flow tanks are restricted if a single continuous weir is used and, therefore, the weir overflow rate can be increased by using a castellated weir or one made up of a series of V-notches. Also, double-sided weirs can be used (Fig. 2.18), although care must be taken about their positioning. If they are located too close to the tank wall, the upflow velocity between the outer weir and the tank wall will be greater than that between the inner weir and the stilling chamber (Boon and Dolan, 1995). Also, their use should not be carried to extremes (Fig. 2.19). Weirs and their overflow channels do need to be kept clean as failure to do this (Fig. 2.20) may affect performance.

Wastewater treatment and technology

Fig. 2.19. Excessive use of multiple weirs

Fig. 2.20. Unwanted growth on weirs/overflow channels

Hydro-dynamic separators

These depend on the development of a vortex in what is, essentially, a cylindrical vessel with a sloping base (Andoh et al., 1993). The liquid flow is introduced tangentially into the side of the vessel at the top of the cylinder. This, together with the drag from the side walls, causes the contents to rotate about the vertical axis creating a vortex which allows solids to settle out. Internal components direct the main fluid flow away from the perimeter into the centre of the vessel as an upward spiralling column rotating slower than the outer downward flow. This difference in rotational speeds further aids the separation of solids so that by the time the flow is discharged from the top of the vessel a high removal of solids has been effected.

This type of separator has been used for grit removal (Gardner and Deamer, 1996), the separation of solids from storm water (Weiss and Michelbach, 1996) and as an alternative to primary settlement tanks (Andoh and Smisson, 1996). Work by Annachhatre and Bhargava (1999) has shown that, when being used as an alternative to primary settlement tanks, the performance was flow related, with the optimum efficiency being achieved at about $5.5-6.0 \, m^3/h$. This gave an SS removal of about 45% and a COD removal of about 40%. The results reported by Andoh and Smisson (1996) showed similar removals, but demonstrated that there would be different removal efficiencies with different types of sewage. They also showed that the performance of a vortex separator could be enhanced by the use of chemicals which aid flocculation.

Sea outfalls

Discharges to coastal waters or estuarine waters are acceptable provided that an element of treatment has been provided. In the past many discharges of inadequately treated sewage were made through outfalls which were also inadequate. This resulted in beaches being contaminated with sewage solids and the waters with potentially harmful bacteria.

In considering discharges into coastal or estuarine waters, a number of European Directives may need to be considered. The Urban Wastewater Treatment (UWWT) Directive specifies the degree of treatment which should be used depending on whether the waters are deemed to be more sensitive, less sensitive or normal (Chapter 1). The treatments which are necessary for normal (Chapters 3–5) and more sensitive waters (Chapter 6) are discussed elsewhere. Discharges to coastal

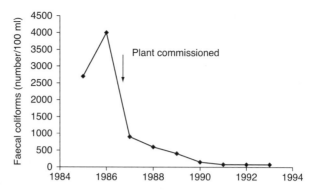

Fig. 2.21. Effect on bacteriological water quality of installing primary treatment and a long sea outfall (Cutting and Setterfield, 1995)

waters which are deemed to be less sensitive require preliminary only or preliminary and primary treatment. The Bathing Waters Directive requires that waters used for bathing should be of a specified bacteriological standard. This may necessitate some form of disinfection (Chapter 7). The Directives which relate to dangerous substances and fisheries may also need to be considered. Installing processes which conform with the UWWT Directive requirements may also result in achieving significantly improved water quality provided that the outfall is of an adequate length and is located so that tidal currents take the discharge away from the beach. The impact on water quality of installing primary treatment and a long sea outfall can be seen in Fig. 2.21. The proper location of an outfall means that a sound understanding of tide levels and currents is essential and this may necessitate the use of floats, radar and tracers (Caine, 1995). Tracers, such as Rhodamine WT, bacterial spores or bacteriophage, will also assist in understanding the degree of dispersion which could be expected.

The criteria used for the design of outfalls carrying treated or partially treated sewage would be

- no gross solids on the beaches;
- an acceptable water quality defined in terms of the Water Quality objectives;
- no accumulation of pathogens in the fish;
- no dissolved oxygen sag;
- no eutrophication;
- no visible slick — this probably means that an initial dilution of 100–400 would be required;
- no localized accumulation ('pockets') in the water or on the bed.

Even with an outfall whose location has been determined by adequate current and dispersion measurement and which complies with the design criteria listed above, it would be advisable to instigate post-commissioning surveys. These would examine the ecology and chemistry of the sediments in the area and the ecology of the seabed (Nixon, 1991).

References

Andoh, R. Y. G., Fagan, G. and Cook, R. (1993) The Swirl-Flo^{TM} physico-chemical treatment process for urban wastewater treatment, *IWEM Symp. on Advances in Wastewater Treatment for Smaller Communities*, Weston-super-Mare, Nov.

Andoh, R. Y. G. and Smisson, R. P. M. (1996) The practical use of wastewater characteristics in design, *Wat. Sci. Technol.*, **33(9)**, 127–34.

Annachhatre, A. P. and Bhargava, A. (1999) Evaluation of Swirl-Flo^{TM} clarifier and UV/H_2O_2 process for domestic wastewater treatment, *Environ. Technol.*, **20**, 285–92.

Boon, A. G. and Dolan, J. F. (1995) Design of settlement tanks and the use of chemicals to aid precipitation of suspended solids, *J. CIWEM.*, Centenary Issue, 57–68.

Caine R. F. (1995) Sea outfalls for the disposal and treatment of wastewater effluent, *Wat. Sci. Technol.*, **32(7)**, 79–86.

Clay, S., Hodgkinson, A., Upton, J. and Green, M. (1996) Developing acceptable sewage screening practices, *Wat. Sci. Technol.*, **33(12)**, 229–34.

Cutting, W. and Setterfield, G. H. (1995) Operation Seaclean, *Paper presented at the CIWEM Centenary Conference, Showing the Way — Benchmarks in UK Environmental Technology*, London, October.

Gardner, P. and Deamer, A. (1996) An evaluation of methods for assessing the removal efficiency of a grit separation device, *Wat. Sci. Technol.*, **33(9)**, 269–75.

Newman, G. D. (1986) Operational experiences with a screenings washer–dewaterer, bagger and incinerator, *Wat. Pollut. Control*, **85**, 57–62.

Nixon, S. C. (1991) Estuarine and coastal waters: Tenby and Weymouth case studies, *J. IWEM*, **5**, 450–9.

Weiss, G. J. and Michelbach, S. (1996) Vortex separator: Dimensionless properties and calculation of annual separation efficiencies, *Wat. Sci. Technol.*, **33(9)**, 277–84.

3
Trickling filters

Introduction
Trickling filters are the older of the two most common processes for the treatment of municipal wastewater. As can be seen from Table 3.1, which compares the main characteristics of these two methods, it is a process which relies on a biofilm to achieve the oxidation of carbonaceous BOD and the oxidation of ammoniacal-nitrogen. The trickling filter is a packed bed reactor with a down-flow of settled sewage and an upflow of air (Fig. 3.1). The packing, which can be angular stone, slag, gravel or clinker, is laid on drainage tiles which overlay a concrete floor (100–150-mm thick). This has a floor slope of 1:100. Filters can be circular or rectangular and the settled sewage is applied by distributors which are driven by water jets, motors or wires. As a process it can be operated in low-, intermediate- or high-rate modes, with the definition of each class being governed by the organic or hydraulic

Table 3.1. A comparison of the main characteristics of trickling filters and the activated sludge process

Trickling filter	Activated sludge
Uses support media	No support media
Fixed bio-film process	Suspended floc process
Natural aeration	Forced aeration
Complex ecology	Simpler ecology
Full treatment	Full treatment

Trickling filters

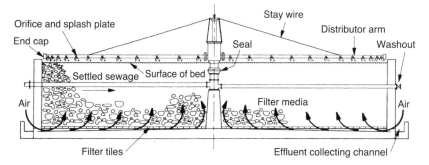

Fig. 3.1. Schematic diagram of a trickling filter (with permission from CIWEM)

loading rate. However, it should be noted that currently in the UK the trend is for it not to be used for large treatment works.

Biofilm

The biofilm which colonizes the packing has a fairly complex mixture of species (Fig. 3.2) ranging from bacteria to protozoa and higher predators (worms and fly larvae). It forms a carefully balanced food web and if this balance is disturbed there can be problems.

The bacteria form the lowest trophic level and do most of the work of purification. The range of bacteria in a biofilm includes both aerobic and anaerobic species, and these could be heterotrophic (organic carbon users) or autotrophic (inorganic users) species. Thus, carbonaceous BOD is oxidized by the heterotrophs, which tend to dominate the film in the upper levels of the bed. The genera typical in these biofilms include *Alcaligenes*, *Flavobacterium* and *Pseudomonas*. Ammoniacal-nitrogen is oxidized by the autotrophs, which tend to populate

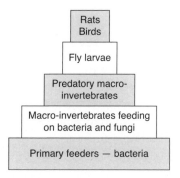

Fig. 3.2. Ecological trophic pyramid for the biofilm of a trickling filter

the lower levels of low-rate filters where the film is well aerated. This stratification is simply due to competition (Wanner and Gujer, 1984). The autotrophs grow more slowly than the heterotrophs and cannot compete successfully in the upper layers where there is a rich supply of carbonaceous material and lower levels of dissolved oxygen. Concentrations of BOD in the liquid film need to be less than 20 mg/litre for nitrification (Parker and Richards, 1986). Consequently, if the applied loading rate increases, as it can do with time, the heterotrophs will gradually become dominant deeper into the filter, and gradually cause nitrification to decrease and, ultimately, cease (Parker and Richards, 1986). Nitrification, the oxidation of ammoniacal-nitrogen to nitrate, is mainly carried out by *Nitrosomonas*, which converts ammoniacal-nitrogen to nitrite-nitrogen, and *Nitrobacter*, which oxidizes nitrite-nitrogen to nitrate-nitrogen (Chapter 6). However, the use of techniques such as fluorescent *in situ* hybridization (FISH) are demonstrating that there are other nitrifying species such as *Nitrosococcus*, *Nitrosospira* and *Nitrospira* (Daims et al., 2001).

The protozoa which exist in biofilms include amoebae, flagellates and ciliates. Their role is, essentially, two-fold

- to feed on any free-swimming bacteria and, thus, maintain the clarity of the effluent;
- to feed on the biofilm and, thus, stimulate bacterial growth.

The metazoan species which can be found in biofilms include rotifers and nematode worms. These are both grazing species which live in and on the film. Lumbricid and enchytraeid worms and fly larvae are among the other species which can be found in trickling filters and which contribute to the control of the biofilm. These species can contribute to the phenomenon of 'spring sloughing' (see later).

In specific circumstances, algae and fungi can be found in trickling filters. Algae would only be expected in the very top layers where there is access to sunlight and they can grow as thick mats resulting in 'ponding' (see later). Fungi, such as *Geotrichum* or *Fusarium*, tend to be dominant when the wastewater being treated has a low pH or is rich in carbohydrates. They grow as filamentous hyphae (collectively called mycelium) and have a faster growth rate than bacteria. This results in a relatively thick, matted film which can, if not controlled, be another cause of 'ponding' (see later).

It should also be recognized that vertebrates such as birds and rats can also have an effect on the filter community by feeding on the growth at the surface.

Trickling filters

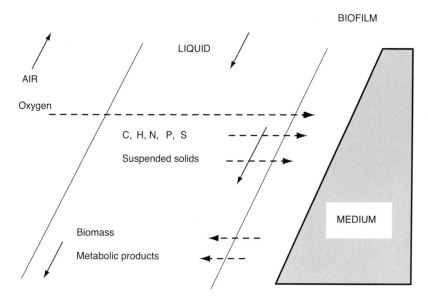

Fig. 3.3. Typical diffusion processes in a trickling filter

The development of the biofilm is, obviously, critical to the operation of a trickling filter. With a new filter, an acceptable effluent will be obtained after a period of several weeks to several months. However, it will probably take two years for the film to reach full maturity. If possible, start-up should be done during the summer months so that there is the opportunity for colonization by grazers before the winter months. It must be remembered that it is the grazing fauna which control the film thickness to the ideal value of 1–2 mm. Therefore, a filter having a film with an inadequate population of grazers is at risk of ponding (see later).

The operation of the biofilm is controlled by diffusion. The settled sewage passing down through the bed forms a liquid film overlying the biofilm. Oxygen enters this liquid film by diffusion. The dissolved oxygen and other nutrients in the liquid film are transferred to the biofilm by a further diffusion process (Fig. 3.3). Similarly, metabolic end-products are released from the biofilm by diffusion.

Media
The packing material in a trickling filter
- must have a high specific surface (m^2/m^3 of filter material) so that the area available for colonization by the biofilm is maximized;

Table 3.2. Characteristics of typical mineral media

	62 mm clinker	50 mm rock	100 mm slag
Bulk density — dry (kg/m^3)	1000–1100	1350	1080
Specific surface area (m^2/m^3)	120	105	50
Voidage (%)	45–50	50	49

- must have a high voidage so that there is an adequate space for the downward movement of liquid and the upward movement of air;
- should be inert and durable and be able to withstand weathering;
- should have mechanical properties such that it is resistant to abrasion during transportation and handling.

In the UK these essential properties are defined in BS 1438: 1971 *Media for Biological Percolating Filters*. Traditionally, in low-rate filters, mineral material has been used for the packing of trickling filters. The types of material which have been used include angular stone, gravel, clinker and blast furnace slags. Their main properties are given in Table 3.2.

The size of media which is used varies from country to country. In the UK, the practice has been to use 100–150 mm material at the bottom (0·3 m) of a filter and then to use 35–50 mm material for the remainder of the bed. In the USA and Germany the practice is to use a uniformly sized bed of 60–100 mm and 40–80 mm respectively. The depth of the bed also varies from country to country, but in the UK it is usual for the bed depth to be 1·8 m unless forced ventilation is used.

The media is laid on some facility which will permit the clear drainage of the treated effluent and the inflow of air. This can take the form of a false floor of perforated tiles or half-round field drains laid in a herring-bone pattern.

Ventilation

In general, natural ventilation is used to supply the air necessary for the bio-oxidation processes. A ventilation column, usually in the form of a vertical pipe, permits air to enter the drainage zone at the bottom of the filter and, thence, pass upwards through the bed. If the filters are totally enclosed, to prevent odour or fly nuisance, continuous, mechanical ventilation must be provided.

Distributors

Although there are a variety of different types of distributor for circular filters, the rotary distributor is probably the most widely used (Fig. 3.4). This consists of four (usually) tubular distribution arms attached to a rotating column at the centre of the filter. The outlets from these arms can be simple holes, spray nozzles or jets and the energy of the liquid coming out of these holes provides the drive for the distributors. The spacing between the holes is varied, being closer the further away from the centre, so that there is a uniform rate of application per unit area. The loss of head for this type of system is about 0·5–1·0 m. Rotary distributors can be driven electrically if it is considered that it is necessary to have a rate of rotation which is independent of the flow-rate of settled sewage. The frequency at which a filter is dosed will affect both the ecology of the biofilm and the performance of the filter. The optimum frequency of dosing will depend on the nature of the wastewater, but a typical value would be between 6 and 30 minutes for every revolution of the distributor. The flow to distributors, which are usually driven by jet reaction, is controlled by a dosing siphon. Its capacity is critical (Roberts and Sambridge, 2000). If it is too large the media can dry out at night when flows are low. If it is too small, the discharge will be too small to drive the distributors effectively. The recommended size is 3 to 4 litres per square metre of filter bed (Pike,

Fig. 3.4. Circular trickling filter (courtesy of Severn Trent Water)

Fig. 3.5. Rectangular filter

1978). Regular maintenance of both the siphon and the distributors is necessary to ensure good performance of the filters.

Reciprocating distributors are the most common for rectangular filters. These span the width of a filter and travel backwards and forwards down the bed, running on the walls of the filter as a bridge or cantilever (Fig. 3.5). They are driven by a continuous rope drive or by water wheels. With the continuous rope drive it is essential that the wheels run on rails to minimize any 'crabbing', which could cause breakage of the ropes. If this occurs, the filter is out of action for a period of time with the possibility of a deterioration of plant performance. Reciprocating distributors are fed by a siphon from a channel running the length of the filter. This channel could be common to two adjacent filters. The outlets from this type of distributor can be holes/nozzles, as with the rotary distributor, or, if the distributor is not tubular, V-notches.

Operation of trickling filters

Modes of operation
There are various configurations for operation. The simplest is single pass filtration (Fig. 3.6). The settled sewage passes through the filter,

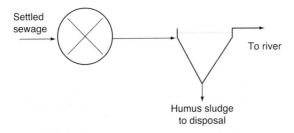

Fig. 3.6. Schematic diagram of single pass filtration

into the humus (final settling) tank (see next section) and, thence, to tertiary treatment (Chapter 7) or the receiving watercourse. Because it is a film process, it is not as easy to determine the residence time in the filter as it is for an open tank. It will depend on a number of factors: the size and shape of the media, the film accumulation and the hydraulic loading rate; but, usually, it would be expected that the liquors would have been in contact with the biofilm for 30–60 minutes. If it is necessary to know a precise value, tracer studies can be used. Although radioactive tracers are the best, it is more common to use inorganic compounds, such as lithium chloride or sodium chloride, or microbiological tracers.

The design of trickling filters is based on the organic loading rate (OLR) which has the units of $kg\,BOD/m^3$ of filter material per day. The OLR is calculated from

$$(\text{BOD concentration} \times \text{flow-rate})/\text{volume of media} \qquad (3.1)$$

For single pass filters, the design OLR is between 0·06 and 0·12 kg/m^3 per day. The hydraulic loading rate (m^3 of flow/m^3 of filter material per day) is also important and, as can be seen from Fig. 3.7, will influence the effluent quality.

Thus, a more refined view of loading rates would be based on the operational requirements, as shown in Table 3.3.

Filters can also be operated with re-circulation (Fig. 3.8). This involves returning a proportion of the treated effluent and mixing it with the incoming settled sewage. The advantage of doing this is that it dilutes the incoming waste and will, thus, reduce the growth of biofilm on the uppermost layers of the filter. This is particularly useful when there is a significant contribution of strong industrial waste in the sewage. With re-circulation, the acceptable OLR is 0·09–0·15. Re-circulation will also minimize the risk of 'ponding' (see later) as filters age or become overloaded.

Wastewater treatment and technology

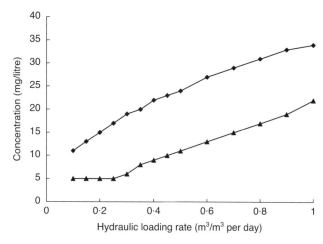

Fig. 3.7. Effect of varying the hydraulic loading rate on the effluent quality (♦ = BOD; ▲ = ammoniacal-nitrogen) from a stone (50 mm) packed filter

An alternative way of dealing with strong sewages (i.e. settled sewage BOD > 300 mg/litre) is to use double filtration. This can be achieved as simple double filtration (Fig. 3.9) or as alternating double filtration (Fig. 3.10). With the former, the media used in the primary filter would be larger than that used either in the secondary filter or in single pass filtration. Typical combinations are 65 mm followed by 28 mm or 100 mm, followed by 40 mm. The primary filter would remove the majority of the BOD load, leaving the secondary filter to remove the residual BOD and achieve nitrification. The total volume of media required would be based on the hydraulic loading rate (m^3/m^3 per day), and this would depend on the standard of effluent required and the size of media being used (Fig. 3.11). Having derived the filter volume the organic loading rate on the primary filter would need to be compared with the maximum accepted value for the size of media, for example, 0·25 kg BOD/m^3 per day for 65 mm stone.

Table 3.3. Typical loading rates in relation to the performance required

Mode	Hydraulic load (m^3/m^3 per day)		OLR (kg/m^3 per day)	
Performance (BOD:SS)	20:30	+Nitrification	20:30	+Nitrification
Single	0·45	0·35	0·12	0·08
Re-circulation	0·60	0·40	0·15	0·10

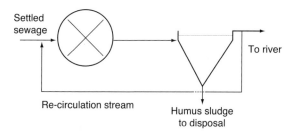

Fig. 3.8. Schematic diagram of filtration with re-circulation

Alternating double filtration (ADF) can also handle loads up to 0·25 kg BOD/m^3 per day. In Phase 1 (Fig. 3.10), the primary filter (A) takes the high initial load and starts to accumulate a heavy biofilm. The secondary filter (B) receives the residual load, resulting in a relatively sparse biofilm. When the flow is changed (Phase 2), filter B has the capacity for the biofilm accumulation which arises from the high load. Filter A will experience quasi-starvation conditions and appreciable amounts of the food storage materials in the biofilm will be utilized to support the gradually diminishing populations of the components in the food web. These processes are repeated with each alternation of flow. In other words, ADF is a process which enables the engineer to manipulate the biofilm when high loading rates are needed. However, the frequency of alternation is site-specific and will depend on the strength and composition of the sewage which is being treated. The

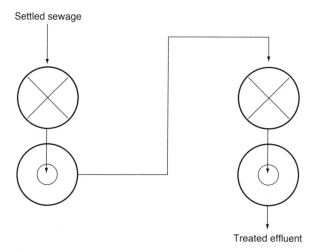

Fig. 3.9. Schematic diagram of simple double filtration

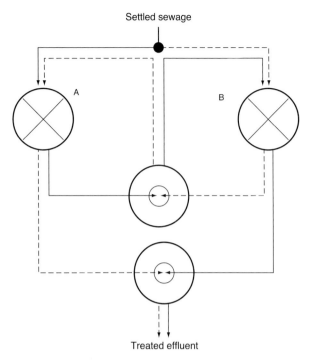

Fig. 3.10. Schematic diagram of alternating double filtration: (——— Phase 1), (– – – – Phase 2)

process control needed to achieve consistently high-quality effluents, together with the higher pumping costs, have made this a less popular option than it was several years ago. It should also be noted that if ADF is to be used, the interstage settlement tank should be designed

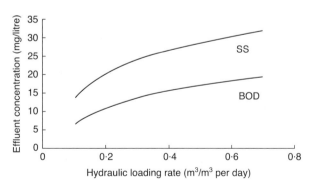

Fig. 3.11. Hydraulic loading rate in relation to effluent quality for double filtration using 65:28 mm media

Table 3.4. Typical overflow rates for humus tanks

Operational mode	Upflow velocity (m/h)		
	DWF	1·2 DWF	Maximum flow
Single pass	0·50	0·60	1·5
Re-circulation	0·66	0·73	1·00
ADF-primary	1·33	1·60	
ADF-secondary	0·66	0·80	1·33

with an overflow rate which is different from that used for the final settlement tank.

Settlement tanks

The settlement characteristics of the solids discharged from trickling filters have not been examined in anything like the detail that activated sludge solids have been. As a result, there are no reliable theoretical mathematics which can be used as a basis for design. The final settlement (humus) tanks are, therefore, designed on the arbitrary basis of values for the upflow velocity in m/s, i.e. (Q/A), where Q is the maximum flow rate (m^3/s) and A is the plan area (m^2) of the settlement tanks. Typical values are shown in Table 3.4, but it must be noted that, whilst some engineers agree with the value of 1·5 m/h for the upflow velocity at maximum flow for single pass filtration, others recommend a value of 1·2 m/h.

Performance

Spring sloughing (see later) can cause the quality of trickling filter effluents to deteriorate during this period, but, in general, they can oxidize BOD and ammoniacal-nitrogen very effectively. An examination of the annual mean data for effluents from some 90 trickling filter plants in a single region showed that 58% of them had ammoniacal-nitrogen concentrations of <5 mg/litre and 60% had BOD concentrations of <15 mg/litre. The influence of factors like spring sloughing showed in the suspended solids' concentrations, with less than 50% of the works achieving values of <25 mg/litre.

The removal of pathogens and parasites is generally erratic and can vary from 20% to 90%. This is also true for the removal of enteroviruses

(Moore et al., 1981). It has been noted recently that there was no significant difference between trickling filters and the activated sludge process for the removal of *Giardia* cysts and *Cryptosporidium* oocysts, although the removal of the former was greater in both types of plant (Robertson et al., 2000).

Operational problems

Ponding

As a trickling filter ages it is likely that the stone media at the surface will decrease in size due to constant weathering. This means that there will be a closer packing of the media and a lower voidage. In addition, the applied load will probably be appreciably greater than that for which the plant was designed, resulting in the formation of a thick biofilm in the region of the surface. The net effect is that there will be some areas of the surface which become clogged and do not allow the ready movement of settled sewage down into the main part of the bed. This is known as ponding (Fig. 3.12). It can also be caused by the accumulation of fine solids, which can be absorbed into the biofilm, but in the winter months are not metabolized, by

Fig. 3.12. Ponding on a trickling filter

the accumulation of gross solids, such as leaves, or by the excessive growth of weeds. Damaged media can and should be replaced. The problem of gross solids ought to be dealt with by good husbandry and plant management. The problem of fine solids is less easily handled and difficult to avoid. The only real solution is to minimize the impact by examining the use of low-frequency dosing and recirculation. Overloading requires an extension of the secondary treatment capacity. If the problem is predominantly the organic load, it is worth considering the removal of some of the load with a high-rate process prior to the main trickling filter (see the next section). The growth of moss on the surface of filters may also lead to ponding. If moss does start to cause a problem, it can be removed by forking off excessive growth or by burning it off with a flame gun. The problem of weed growth is relatively uncommon, but where weeds do thrive and cause problems the only real solution is their removal, either mechanically or manually, as there are no herbicides which have the approval of the Health and Safety Executive for use on filters.

Flies

The flies, which are an integral part of most trickling filter ecologies and, therefore, emerge from filters, belong to three families of Diptera:

- Psychodidae (the filter fly);
- Chironomidae;
- Anisopodidae.

These are not biting species and are not a public health risk but, in large numbers, are a nuisance and an environmental risk, and fly control becomes critical if habitable buildings are close. Indeed, under the Environmental Protection Act of 1990 (Section 80), abatement notices can be served on water companies requiring that they prevent public nuisance. The size of the problem can be judged from the number of flies which can swarm from a filter, up to 20 000–50 000/ m^2 per day (Edeline, 1988; Roberts and Sambridge, 2000). In considering a control policy, it is important to know which type of fly is causing the problem; the filter fly 'drifts' and will move in the direction of the prevailing wind. The other species fly so that their movement is independent of the wind direction. It may be possible to identify 'resting points' for the fly swarms and to use insecticides there, but it is more likely that control will have to be exercised within the filter.

The type of biofilm within the filter will to some extent determine which species will be dominant. Psychodidae tend to be found in filters where there is a thick film and Chironomidae where there is a thin film. One control route, therefore, is to create inter-species competition so that no one species is dominant. In this way the overall numbers are reduced. Low-frequency dosing and the use of re-circulation are ways by which this can be achieved.

The use of plastic netting with a pore size of about 2 mm over the surface of trickling filters provides a barrier, which prevents flies from leaving the filter but which allows the passage of the settled sewage into the filter. Its barrier effect is enhanced by the film of liquid which is held across the pores by surface tension. This is a technique which has been proved to be effective and has little environmental impact.

The use of insecticides within the filter should be considered as a last resort. Only insecticides which are approved by the Health and Safety Executive may be used and constraints may be imposed to protect the receiving watercourses. The insecticides which have been approved contain either a bacterial toxin, which has to be ingested by the fly larvae to be effective, or an insect growth regulator.

Spring sloughing

Spring sloughing is the result of the effect of temperature on the different species in the biofilm. During the winter, the activity and numbers of the higher, predatory species is reduced. The variation in the numbers of grazing animals throughout the year can be seen in Fig. 3.13. This means that the bacterial component of the film and the film density increase. In spring, as the temperature increases, the predators become active again and this results in the discharge of 'lumps' of biofilm. This can overload settlement tanks with the resultant discharge of an effluent with higher than normal concentrations of suspended solids (Fig. 3.14).

Temperature

Low temperatures during the winter months will both affect the ecology of the biofilm and reduce the metabolic activity of all the species. The autotrophic species which carry out nitrification are the most susceptible to cold weather. The result of this can be higher ammoniacal-nitrogen concentrations in the treated effluent during winter

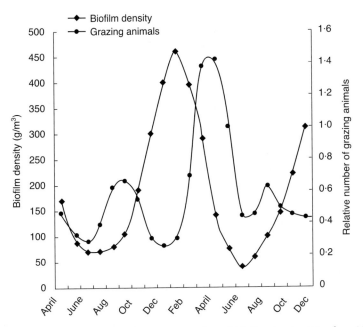

Fig. 3.13. *Seasonal variation of grazing animals and biofilm density (Hawkes, 1963)*

months, as can be seen in Fig. 3.15. The use of artificial heating has been examined, but it was concluded that its use to increase the capacity of trickling filters in cold weather could not be recommended (Gebert and Wilderer, 2000). However, if filters are located in very exposed

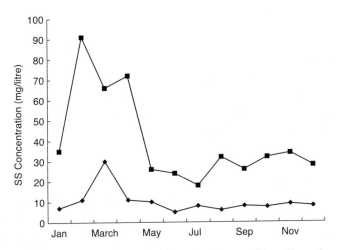

Fig. 3.14. *Mean monthly SS concentrations from two trickling filters showing the occurrence of spring sloughing*

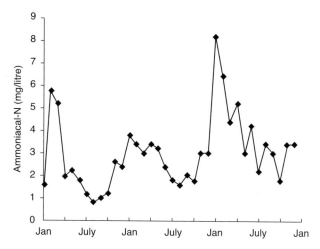

Fig. 3.15. Final effluent quality from a trickling filter required to meet an ammoniacal-nitrogen standard of 5 mg/litre

areas, some form of screening might be considered to minimize wind-chill effects which would increase the effect of lower temperatures.

High-rate filters

Mineral media can be used for high-rate filters, the aim of which is to achieve only a partial removal of BOD. This type of filter uses a coarse grade of media, perhaps 100 mm, to minimize clogging and receives loads in excess of 0·5 kg BOD/m^3 per day. It is not a process which is widely used in the UK. At these loadings, the growth of the biofilm is greater than in low-rate filters which means that its residence time is shorter. In turn, this means that the solids which are discharged from the filter are less well oxidized and are, therefore, less stable. These sludges are also difficult to de-water.

Plastic media

High-rate filtration

Filters can also be packed with a plastic media, which can be in the form of modules or a random packing (Figs 3.16 and 3.17). In an evaluation of a range of different types of media, cross flow media had much higher hydraulic retention times than vertical flow or random media and gave better BOD removals (Richards and Reinhart, 1986). Plastic media is

Trickling filters

Fig. 3.16. MARPAK™ *modular plastic media (courtesy of Marley Davenport Ltd)*

Fig. 3.17. *Random packing plastic media*

Wastewater treatment and technology

Table 3.5. *Characteristics of examples of plastic media*

Media	Plastic	Weight (kg/m^3)	Surface area (m^2/m^3)	Voidage (%)
BIOdek 120·60	PVC	40	240	96
BIOdek 190·60	PVC	35	150	97
Flocor Rm	PVC	50	300	97
Flocor R2	PVC	50	140	97
Flocor E	PVC	45	100	96
Filterpak	Polypropylene	64	118	93
Actifil	Polypropylene	44	102	95

light, has a very high voidage and a specific surface which, usually, is greater than that of stone (Table 3.5).

The lightness means that the filters can be built as towers (up to 10 m without intermediate support). As such, they are a low footprint process. The foundations are usually reinforced concrete and the floors slope 1 in 25 for drainage. There must be air access, usually 0·5–4% of the plan area. The irrigation rates tend to be product specific but might be expected to be about 6 m^3/m^3 per day. Achieving this may necessitate some re-circulation. In practice, hydraulic loading rates may be as high as 50 m^3/m^3 per day.

Typically, the applied organic load is 2–7 kg/m^3 per day and the removal efficiency is 50–75% (Fig. 3.18). The natural ventilation rates are such that there can be a degree of cooling within the

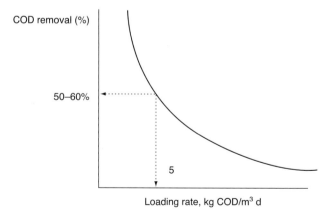

Fig. 3.18. *Generalized performance for a plastic media filter operating at high-rate*

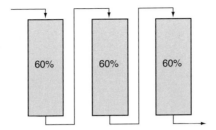

Fig. 3.19. Schematic layout of a cascade sequence with plastic media filters

towers. This can affect performance. The ventilation can create odour problems by stripping out 'smelly' compounds (industrial chemicals or hydrogen sulphide) in the wastewater. The higher organic loads which are applied mean that there is a more rapid growth of biofilm than in low rate filters. Faster growing films tend to be sheared off and flushed away more quickly. The organisms growing under these circumstances are those with short generation times. This means that the sludge discharged from high-rate filters is not well mineralized and will contain appreciable amounts of adsorbed, un-degraded BOD. As such, it will be unstable and has the potential to cause odours during storage. The production of sludge is also appreciable with values of between 0·6 and 1·1 kg SS/kg BOD removed being reported. However, although it is a sludge which does not de-water readily, it does settle well and upflow velocities of $24\,m^3/m^2$ per day are normally used.

Applications of high-rate filters are

- partial treatment of strong industrial effluents. In these cases it may be appropriate to use more than one in a cascade sequence (Fig. 3.19);
- partial treatment of domestic sewage when relaxed consents apply, such as for discharge to coastal or estuarine waters;
- load lopping at overloaded works (see earlier).

Low-rate filtration

There is an increasing use of plastic media as the support material for filters which are required to produce high-quality effluents or which are being used for tertiary nitrification. In some cases these are installed as new filters, in others the plastic media are used to replace the mineral media in existing filters. In the latter case, it may be necessary to increase the depth of the filters. The loading rates which would be used for a filter oxidizing BOD and ammonia would depend on the

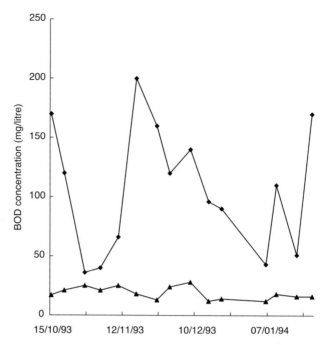

Fig. 3.20. Performance of a trickling filter packed with plastic media operating at low-rate (influent ◆; effluent ▲)

type of media and the effluent quality required, but would be slightly greater than those used for a single pass mineral media filter. For example, the data shown in Fig. 3.20 were taken from a filter with a mean loading rate of 0·165 kg BOD/m³ per day.

Models

A number of models for the prediction of the performance of trickling filters have been proposed. These range from computer-based software, such as that in STOAT, to empirical expressions and equations based on first-order decay (Pike, 1978; Logan et al., 1987). One of the most commonly cited expressions is that derived by the US National Research Council (National Research Council, 1946; Gray, 1999):

$$E = \frac{100}{1 + 0\cdot 448(W/VF)^{0\cdot 5}} \quad (3.2)$$

where E = removal efficiency (%), W = applied load (kg BOD/d), V = filter volume (m³), F = re-circulation factor. However, it has been said that 'these equations were not consistent at predicting performance successfully' (Boon et al., 1997). Therefore, there is a need to examine their suitability. Looking at the case where there is no re-circulation and there is a BOD removal of 95%, the National Research Council equation can be re-arranged as

$$\frac{W}{V} = \left[\frac{(100/95) - 1}{0.448}\right]^2 \quad \text{giving } V = \frac{W}{0.0138} \quad (3.3)$$

This is considerably different from the expression derived when an organic loading rate of 0·06 kg BOD/m^3 per day is used, i.e.

$$\frac{W}{V} = 0.06$$

References

Boon, A. G., Hemfrey, J., Boon, K. and Brown, M. (1997) Recent developments in biological filtration of sewage to produce high quality nitrified effluent, *J. CIWEM*, **11**, 393–412.

Daims, H., Purkhold, U., Bjerrum, L., Arnold, E., Wilderer, P. A. and Wagner, M. (2001) Nitrification in sequencing batch reactors: lessons from molecular approaches, *Wat. Sci. Technol.*, **43(3)**, 9–18.

Edeline, F. (1988) *L'Epuration Biologique des Eaux Residuaires: Theorie et Technologie*, CEBEDOC, Liege, Belgium.

Gebert, W. and Wilderer, P. A. (2000) Heating up trickling filters to tackle cold weather conditions, *Wat. Sci. Technol.*, **41(1)**, 163–6.

Gray, N. F. (1999) *Water Technology: An Introduction for Environmental Scientists and Engineers*, Arnold, London.

Hawkes, H. A. (1963) *The Ecology of Wastewater Treatment*, Pergamon Press, Oxford.

Logan, B. E., Hermanowicz, S. W. and Parker, D. S. (1987) Fundamental model for trickling filter process design, *J. Wat. Pollut. Control Fed.*, **59**, 1029–42.

Moore, B. E., Sagik, B. P. and Sorber, C. A. (1981) Viral transport to groundwater at a wastewater land application site, *J. Wat. Pollut. Control Fed.*, **53**, 1492–502.

National Research Council (1946) Sewage treatment in military installations, *Sew. Wks. J.*, **18**, 787–1028.

Parker, D. S. and Richards, T. (1986) Nitrification in trickling filters, *J. Wat. Pollut. Control Fed.*, **58**, 896–902.

Pike, E. B. (1978) The design of percolating filters and rotary contactors, including details of international practice, *Technical Report TR 93*, Water Research Centre, Medmenham.

Richards, T. and Reinhart, D. (1986) Evaluation of plastic media in trickling filters, *J. Wat. Pollut. Control Fed.*, **58**, 774–83.

Roberts, P. F. and Sambridge, N. (2000) *Manuals of British Practice in Water Pollution Control; Biological Filtration and Other Fixed Film Processes*, Chartered Institution of Water and Environmental Management, London.

Robertson, L. J., Paton, C. A., Campbell, A. T., Smith, P. G., Jackson, M. H., Gilmour, R. A., Black, S. E., Stevenson, D. A. and Smith, H. V. (2000) Giardia cysts and Cryptosporidium oocysts at sewage treatment works in Scotland, UK, *Water Res.*, **34**, 2310–22.

Wanner, O. and Gujer, W. (1984) Competition in biofilms, *12th Conf. Int. Assoc. Water Pollut. Res.*, Amsterdam.

4
Activated sludge process

Introduction

The activated sludge process is, arguably, the most common form of treatment for municipal sewage, at least in terms of the volume treated. It consists of an aeration tank and a settlement tank (Fig. 4.1) and the two units must be considered as a single unit for design and operational purposes. In the aeration tank the biomass, the mixed liquor suspended solids (MLSS), is contacted with air and, usually, settled sewage. After a pre-determined time the mixed liquor is passed to a settlement tank where the solids and treated effluent are separated, the thickened sludge is returned to the aeration tank and the treated effluent is either discharged to a watercourse or passes on to further treatment (Chapter 7). Returning the thickened sludge is an essential feature of the process as it can be considered as being a continuous fermentation operating under wash-out conditions and, without the solids' feedback, it would not be possible to maintain a high enough concentration of biomass in the aeration tank to achieve the degree of treatment which is required. The 'activation' of the sludge implies that the biomass

- contains all the enzymes necessary to degrade the waste being treated;
- has a surface with good sorption properties;
- will settle readily under quiescent conditions.

The biomass is essentially a mixture of bacteria and protozoa, together with some nematodes and rotifers. The overall cell biomass can be thought of as aggregates which are integrated into an organized

Wastewater treatment and technology

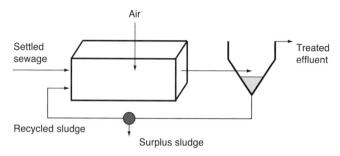

Fig. 4.1. *Schematic diagram of the activated sludge process*

floc. This has a complex structure but it is generally considered to be held together by a polymeric matrix comprised of polysaccharides, proteins, DNA and humic material (Steiner *et al.*, 1976; Horan and Eccles, 1986). It is also considered that there will be interactions between these extracellullar polymers and polyvalent metal ions (Forster and Dallas-Newton, 1980; Morgan and Forster, 1992; Sanin and Vesilind, 2000). The bacteria can be considered in two categories: the floc-formers such as *Achromobacter*, *Pseudomonas* and *Zoogloea* and the filamentous species. At one stage, it was considered that bacteria of the genus *Zoogloea* were the floc-formers which had an essential role in establishing floc structure and stability (Pipes, 1967), but it is now accepted that the filamentous species are an essential factor in floc structure in that they form a 'backbone' for the floc (Sezgin *et al.*, 1978; Jenkins *et al.*, 1993). However, when they grow out from the floc (Fig. 4.2). they create the problems of bulking and foaming

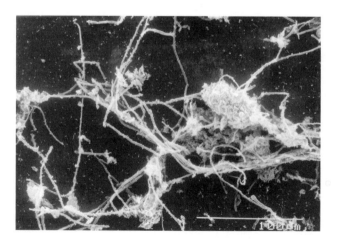

Fig. 4.2. *Activated sludge with a heavy filament outgrowth*

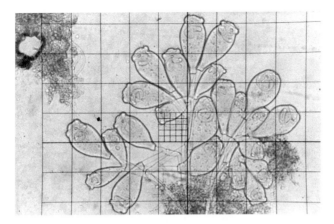

Fig. 4.3. Stalked ciliates attached to an activated sludge floc

which will be discussed later. The protozoan species are also an essential element of the activated sludge in that, because they feed on bacteria, they are responsible for removing turbidity caused by free-swimming bacteria or small bacterial aggregates (Curds et al., 1968). The species which are present will vary with the maturity of the sludge. A relatively immature sludge will contain crawling and free-swimming ciliates, such as *Aspidisca* and *Paramecium*, while a mature sludge will be dominated by stalked ciliates, such as *Vorticella* and *Opercularia* (Fig. 4.3). A very mature sludge will also contain rotifers (Fig. 4.4) and nematodes. This sequential progression in the higher organisms present in the sludge (Fig. 4.5) provides a ready assessment of the nature of the sludge itself (Curds, 1966).

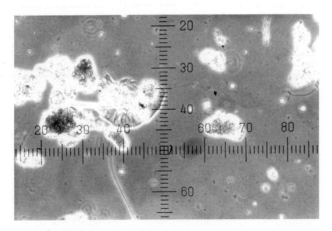

Fig. 4.4. A rotifer grazing on activated sludge flocs

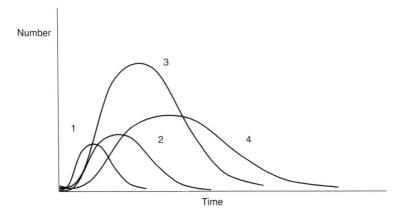

Fig. 4.5. The sequence of microbial dominance in activated sludge showing the relative numbers of flagellates (1), free swimming ciliates (2), bacteria (3) and stalked ciliates (4) in activated sludge

The main design criteria are

- sludge loading rate (food to mass ratio);
- sludge age;
- sludge settleability;
- sludge production.

Process variables

Organic loading rate
The organic loading rate (OLR), also known as the food:mass ratio or the sludge loading rate, is expressed as the mass of BOD/mass of sludge (kg BOD/kg MLSS per day). This can be calculated from

$$\text{OLR} = \frac{B \times Q}{M \times V} \qquad (4.1)$$

where B = BOD of the settled sewage (g/litre), Q = flow-rate of settled sewage (m^3/d), M = mixed liquor suspended solids (g/litre), V = aeration tank volume (m^3). The effluent quality will depend on

- temperature;
- mixed liquor suspended solids;
- sludge loading rate.

These relationships between effluent quality and OLR are shown in Table 4.1. Thus, knowing the effluent quality which is required (i.e.

Table 4.1. *Effluent quality and organic loading rate*

Rating	Effluent quality	OLR
High rate	Non-nitrified; worse than Royal Commission	0·5–5·0
Conventional-1	95 percentile BOD = 20; some nitrification	0·25–0·50
Conventional-2	95 percentile BOD = 10; some nitrification	0·15–0·25
Conventional + nitrification	BOD c. 5–12; amm.N c. 5–10	<0·15
Extended aeration	BOD c. 5 amm.N c. 2·5	0·05–0·10

the consent), the sludge loading rate can be determined. The sludge loading rate will also affect the amount of excess sludge which is produced during bio-oxidation (the growth index). This relationship is given in Fig. 4.6 and shows that, at higher loading rates, more sludge is produced.

Sludge age

This represents the average solids' retention time within the system. It is controlled by the sludge wastage rate which, in turn, is dictated by the sludge yield (or sludge production rate). Having the right sludge age can be of critical importance if the process is required to nitrify (i.e. oxidize ammonia). The bacteria which carry out the nitrification process, *Nitrosomonas* and *Nitrobacter*, grow more slowly than the carbon oxidizers. This means that, unless there is a long enough solids' retention time to permit their growth, it will not be possible to maintain an adequate population of nitrifiers in the mixed liquor. There are equations involving the temperature (T in degrees Celsius) and pH which can be used for calculating the appropriate sludge age. For

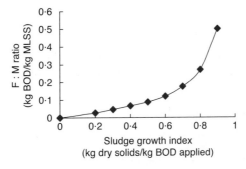

Fig. 4.6. *The effect of loading rate on the growth of activated sludge*

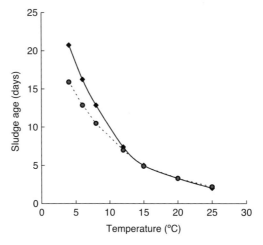

Fig. 4.7. The relationship between temperature and the minimum sludge age for nitrification (● Severn Trent Water, 1995; ▲ Marais, 1973)

example (Downing, 1968),

$$(1/\text{sludge age}) = [0.18 - 0.15(7.2 - \text{pH})]\exp[0.12(T - 15)] \quad (4.2)$$

Alternatively, it can be derived from data similar to that in Fig. 4.7, which shows the minimum sludge age to maintain nitrification at a range of temperatures, or by the 'rule of thumb' which dictates a sludge age of 10 days at 10 °C. The sludge age is derived from the mass of solids under aeration and the daily mass of solids lost from the system. This latter element should include both the solids which are wasted intentionally and those lost in the final effluent. Thus, it is calculated from

$$\text{sludge age} = \frac{MV}{Q_W S_W + Q S_E} \quad (4.3)$$

where M = mixed liquor suspended solids (g/litre), V = aeration tank volume (m³), Q_W = volumetric flow rate of wasted sludge (m³/d), S_W = suspended solids' concentration of wasted sludge (g/litre), Q = flow through plant (m³/d), S_E = effluent suspended solids' concentration (g/litre).

Hydraulic retention time

This is also an important parameter in the operation of the activated sludge plant and can range from about 8 hours for a conventionally

Activated sludge process

operated plant to >24 hours for an extended aeration system. For an activated sludge plant which is required to nitrify, Severn Trent Water use a nominal hydraulic retention time of 9–12 hours (Churchley et al., 2000).

Aeration plant design

Aeration serves a number of purposes. First and foremost it must provide an adequate and continuous supply of dissolved oxygen (DO) for the sludge. It also ensures that the MLSS is kept in suspension and is mixed with the incoming sewage. The metabolic processes which oxidize the BOD generate carbon dioxide. This is an acid gas which is partially soluble in water. The aeration processes strip this carbon dioxide out of the mixed liquors, thus preventing any accumulation and the associated drop in pH.

Aeration devices

Aeration systems must be capable of supplying oxygen at a greater rate than the rate of oxygen utilization by the biomass. Essentially, there are two main types of aeration system: diffused air and surface aerators. Air is normally diffused through ceramic diffusers into the mixed liquor near to the bottom of the tank to produce fine-bubbles (2 to 5 mm diameter) so as to achieve mixing and maximum contact period (Fig. 4.8).

There are two basic types of mechanical surface device, one which has a vertically mounted (cone-type) aerator (Fig. 4.9) and the other a horizontally mounted (brush-type) aerator (Fig. 4.10). The efficiencies of both types vary with their depth of immersion in the mixed liquor. Other types of aeration system which use air do exist and are used; for example, the submerged turbine (Fig. 4.11) or jet aeration (Fig. 4.12). However, they are by no means as common as the fine bubble diffusion and mechanical surface aeration systems.

Tests in the UK of these aeration devices have shown that the fine-bubble diffused-air system can give the greatest aeration intensity (g O_2 supplied/m^3 aeration tank capacity per hour) and efficiency (g O_2 supplied/W h), provided the system is correctly installed, operated and maintained. The change in the dissolved oxygen concentration with time (in hours) is given by

$$\frac{dC}{dt} = (K_L a)(C^* - C) \tag{4.4}$$

Wastewater treatment and technology

Fig. 4.8. Layout of fine-bubble aerator domes

Fig. 4.9. A vertical shaft mechanical surface aerator

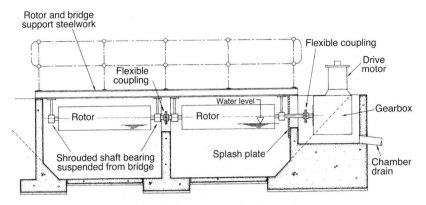

Fig. 4.10(a). The layout of a horizontal shaft mechanical surface aerator

Fig. 4.10(b). The operation of a horizontal shaft mechanical surface aerator

where $K_L a$ = mass transfer coefficient (h^{-1}), C = dissolved oxygen concentration (mg/litre), C^* = saturated dissolved oxygen concentration (mg/litre). Thus

$$\log(C^* - C) = (K_L a)t \tag{4.5}$$

This means that if the DO concentration in an aerated vessel is measured with time, starting with an initial DO concentration of zero (or close to zero), the value of $(K_L a)$ can be determined from a plot of $\log(C^* - C)$ against time. The DO is lowered by the addition of sodium sulphite, together with a cobalt chloride catalyst. The value for $K_L a$ is then corrected to the 'standard' temperature of 20°C by

$$(K_L a)_T = (K_L a)_{20}(1 \cdot 024)^{(T-20)} \tag{4.6}$$

Wastewater treatment and technology

Fig. 4.11. *A submerged turbine aerator*

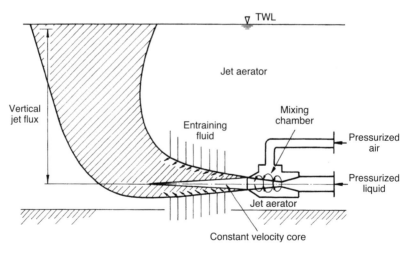

Fig. 4.12. *A schematic diagram of the jet aeration system*

Table 4.2. Values of alpha

Source	α
Badger, Robinson and Kiff, 1975	0·63–0·94
Rees and Skellett, 1974	0·89–0·93
Lister and Boon, 1973	0·3–0·8
Eckenfelder and O'Connor, 1961	0·48–0·86
Tewari and Bewtra, 1982	0·67–0·925

Knowing the aeration tank volume (V) and the electrical wattage used during the aerator trials enables the oxygenation capacity (OC) and the aerator efficiency (E) to be calculated

$$OC = (K_L a)_{20} V C_{20}^* \qquad (4.7)$$

$$E = \frac{(K_L a)_{20} \times V \times C_{20}^* \times 10^{-3}}{kW} \quad \text{in } kg\,O_2/kWh \qquad (4.8)$$

where 1 kilowatt-hour (kWh) = 3·6 MJ. This is clean water efficiency quoted by manufacturers. However, many of the components in sewage will effect the values of both C^* and $K_L a$. Essentially, the concern is with surface active molecules in the sewage, which can change water from a coalescing fluid into a non-coalescing one. Two proportionality factors, α and β, are, therefore, used

$$\alpha = \frac{K_L a \text{ in sewage}}{K_L a \text{ in water}} \qquad (4.9)$$

$$\beta = \frac{C^* \text{ in sewage}}{C^* \text{ in water}} \qquad (4.10)$$

Beta is relatively constant with a value of about 0·9. The value of alpha, however, can be very variable (Table 4.2). For example, experiments with fine bubble diffusors have shown that, with an essentially plug-flow system, the value of alpha was 0·3 at the start of treatment and was 0·8 at the end of the aeration tank when a high quality, fully nitrified effluent had been obtained.

Aerator efficiencies

'Alpha' values do not indicate the relative efficiency of aeration systems. A fine-bubble diffused-air (FBDA) system may have an 'alpha' value as low as 0·4 at the inlet end of a plug-flow aeration unit, and in a mechanically operated system it may be as high as 1·2. The relative efficiencies of different aerators can be judged only by

Fig. 4.13. A vertical shaft mechanical surface aerator with a draft tube

comparison of their mass-transfer coefficients, aeration intensities or aeration efficiencies when tested under similar conditions in aerated liquor (or in tap water or final effluent with or without added surfactant). For example, a fine-bubble aeration system may have an aeration efficiency of 2 kg/kWh at the inlet of a plug-flow aeration tank and 4 kg/kWh at the outlet, whereas a mechanical surface aerator could have corresponding values of 1·8 and 1·5 kg/kWh respectively. In clean water, the FBDA systems may have an efficiency of about 6 g/W h. There is also a need to consider whether or not to use draft tubes with vertical shaft mechanical aerators (Fig. 4.13). It is claimed that the use of draft tubes does give a greater efficiency. In comparing the relative merits of the two types of aerator, it should be noted that there is a potential risk of aerosol formation with FBDA systems. Mechanical surface aerators do not have this problem.

Aerator efficiencies do need to be checked during the commissioning of a plant and it may be necessary to re-examine them as time progresses to determine whether there has been any deterioration in the performance. The procedure for deriving $K_L a$ and the oxygenation capacity does have some potentially awkward aspects, and the following points need to be considered

- How much cobalt chloride catalyst should be used? 0·5 mg/litre is the specified amount, but how much is adsorbed onto the concrete walls of the tank? The final concentration should be measured;
- How many trials should be done and what are the effects of re-using the water?

- If the efficiency is being assessed against a manufacturer's specification, does it matter whether the tank dimensions are different from those used by the manufacturer?
- What constitutes a failure? A difference of 50% is clear-cut but would a 10% difference be considered to be a failure?
- What variability is there in the measured efficiency? Data from multiple trials suggest that one of ±5% might be expected.

Oxygen transferred by aerators

In considering how much oxygen is transferred by an aerator, the efficiency, the values of α, β, the operating dissolved oxygen concentration and the temperature need to be considered.

Thus, the transfer of oxygen by a mechanical aerator in terms of kgO_2/kWh is given by

$$\text{oxygen transfer} = \text{aerator efficiency} \times \alpha \times \left(\frac{\beta C^* - C}{9 \cdot 2}\right)$$
$$\times 1 \cdot 024^{(T-20)} \qquad (4.11)$$

The electrical size of the aerator can then be derived from

$$\text{aerator size (in kW)} = \frac{\text{oxygen requirement (in kg/h)}}{\text{oxygen transferred (in kg/kWh)}} \qquad (4.12)$$

Oxygen requirements

The next step is to determine the total oxygen requirement (R). This depends on the efficiency of treatment required and, therefore, the processes taking place within the aeration tank. These can be

- oxidation of carbonaceous BOD;
- endogenous respiration of the biomass (which includes a temperature term);
- oxidation of ammoniacal-nitrogen;
- de-nitrification.

Thus, the total oxygen demand (Johnstone et al., 1983) may be calculated from

$$R = 0 \cdot 75Q(B_s - B_e) + 2 \times 10^{-3}MV + 4 \cdot 3Q(A_s - A_e)$$
$$- 2 \cdot 83Q[(A_s - A_e) - N_e] \qquad (4.13)$$

where R = oxygen requirement (g/h), Q = flow-rate of settled sewage (m^3/h), B_s = BOD of settled sewage (mg/litre = g/m^3), B_e = BOD of final effluent (mg/litre = g/m^3), A_s = amm.N of settled sewage (mg/litre = g/m^3), A_e = amm.N of final effluent (mg/litre = g/m^3), N_e = nitrate-N of final effluent (mg/litre = g/m^3), M = mixed liquor suspended solids (mg/litre = g/m^3), V = aeration tank volume (m^3).

Strictly speaking, this equation is only valid for 10 °C, since the value 2×10^{-3} is derived from the term $0{\cdot}001 \times r_{20} \times 1{\cdot}07^{(T-20)}$. Therefore, if greater precision is required or if the mean operating temperature is significantly different from 10 °C, the specific components should be used

- r_{20} = endogenous respiration rate of the MLSS = 3·9 mg/g h;
- T = maximum temperature since this gives lowest DO;
- factorize R by 1·56 to allow for simultaneous peaking of load (1·2) and flow (1·3).

The components of the equation for deriving the oxygen requirement that are used for any particular situation will only be those which relate to the process being designed. Thus, if de-nitrification is not being applied, only the first three terms will be used. Also, if raw sewage is being fed to the aeration tank (as might be the case for an oxidation ditch), the BOD factor should be 1·0 rather than 0·75.

The value of N_e will depend on the rate of sludge recycling and the degree of de-nitrification which is achieved. In particular, since the rate of sludge recycling (Q_u) is likely to be in the range 0·5Q to 1·5Q, complete de-nitrification can be assumed and the concentration of nitrate in the final effluent will be given by

$$N_e = [A_s - A_e]\left[\frac{Q}{Q_u + Q}\right] \quad (4.14)$$

Optimization of aeration efficiency
Tests of aeration systems have shown that, potentially, the fine-bubble diffused-air system could use less electrical energy to dissolve a given weight of oxygen when compared with most other systems, particularly coarse-bubble aeration. It is also possible to achieve a more efficient use of the oxygen supplied to a plug-flow tank by designing the plant with tapered aeration (Fig. 4.14). By supplying more air at the inlet, where there is the greatest BOD load, and less air towards the tank outlet, where the BOD is practically at the final effluent concentration, it is

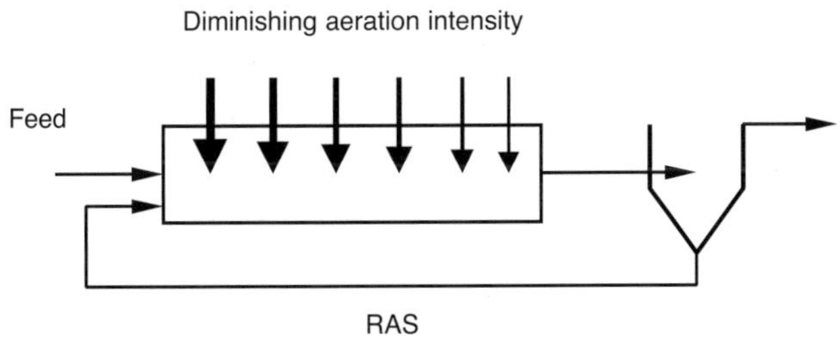

Fig. 4.14. *A schematic diagram showing tapered aeration in a plug-flow aeration tank*

possible to avoid under- or over-aeration at the inlet and outlet respectively. As a 'rule of thumb', if the aeration tank is thought of as five discrete units, the oxygen requirements for each unit along the tank would be 60, 15, 10, 10 and 5% respectively.

As a result of studies carried out by the Water Research Centre (WRc) to evaluate a fine-bubble aeration system and further work undertaken by the USEPA in co-operation with the WRc, Thames Water Authority, Severn Trent Water Authority and others, the following factors have been determined as important in order to obtain optimum performance (lowest energy input) (Chambers and Jones, 1985)

(a) The geometry of the aeration basin and the configuration of the diffusers are probably the most significant factors affecting aeration efficiency. It is recognized that there are advantages to plug-flow of liquor along the aeration tank, particularly in relation to improved settleability of activated sludge, but a length-to-width ratio of about 12:1 could give optimal aeration efficiency. Tapering of aeration intensity by varying the configuration of diffusers along the tank should be provided but not to such an extent that, at times of low BOD and ammoniacal nitrogen load, the required flow-rate of air is below the minimum specified by the manufacturers or the intensity of mixing is inadequate to keep the activated sludge in suspension. The optimum depth of liquor in the aeration tank is probably between 4·5 and 6 m.
(b) Excessively high concentrations of DO have little benefit to the treatment process but significantly reduce the aeration efficiency. Monitoring and control of DO by varying the air flow-rate should result in an increase in aeration efficiency and a saving of energy. To achieve effective control of DO, a minimum of three independently controlled air-grids, each with air-metering and linked to a

DO probe via a programmable controller, would be required. The control system could be provided by a mini-computer capable of varying the output of air supplied by a variable number of blowers, one of which could be driven by a speed-controlled d.c. electric motor.

(c) The intensity of aeration should correspond to the requirement for oxygen. Changes in temperature and treatability of wastewater will affect the rate of biochemical oxidation so that the maximum intensity of aeration should be based on the maximum rate of carbonaceous and nitrogenous oxidation anticipated, preferably as determined from experience rather than based on rigid application of theoretical data.

Aeration control can save electrical costs. The system is based around an oxygen electrode. With FBDA, the control is focused on the blowers. With surface aeration the degree of submersion is controlled. There is also a need for good 'turn-down' (and turn-up) facilities on the aerators to accommodate the fluctuations in load over any 24 hour period. A useful 'rule of thumb' is to install an aeration plant to satisfy between 0·5 and 1·5 times the average hourly demand.

Dissolved oxygen control

Dissolved oxygen control is a very necessary aspect of activated sludge operation. For carbonaceous BOD removal the DO residual should be 0·5–1·0 mg/litre and for nitrification it should be more than 2·0 mg/litre. However, some thought needs to be given to where the controlling electrode should be placed and at what depth.

Looking at respiration, it could be argued that for a plant operating only with carbonaceous BOD removal, control should be located at the front of the tank, and that for C and N removal it should be at the end. Certainly, with a completely mixed tank the control electrode would most probably be located near the outlet weir. Consideration also needs to be given to the frequency of electrode calibration.

A more rigorous analysis suggests that control procedures may depend on the type of aeration system being used.

Fine-bubble diffused-aeration

At larger works, centrifugal blowers with modulating inlet and outlet vanes are used. These give high turn-down capabilities whilst

Activated sludge process

maintaining high efficiencies. They give constant pressure irrespective of the demand. On smaller plants positive displacement blowers would be used. Their speeds can be adjusted to vary the output, but large turn-down ratios are not possible.

The load and, therefore, the oxygen demand will vary with time, and in a tank with significant plug-flow the situation could arise where the local demand was increasing in one part of the tank and decreasing in another. So, for plug-flow tanks, blower control alone is not highly effective. Consideration should be given to 'zone aeration', i.e. to have valves controlling air flow to individual zones and a throttling facility on the valves.

Mechanical surface aeration
There are a number of choices for control

(a) Switch aerators on/off. This is a simple strategy. It is also a cheap strategy. However, in practice there may be only two or three pockets in an aeration lane so that switching off an aerator would remove up to 50% of the installed capacity. Therefore, this should only be considered if there were more than four pockets in a lane. Also, if the DO electrode was in the pocket which had been turned off, its readings would be unreliable.
(b) Vary the immersion by altering the liquid level. In effect, this means varying the final weir setting. This will affect the liquid level in the entire lane and may not be appropriate in a high plug-flow system (see the earlier comments about zone control).
(c) Vary the immersion by altering the aerator level. It works, but the pay-back is not favourable.
(d) Vary the aerator speed. This will require a control loop for each motor, but it may be the best option.

A more detailed break-down of the options is given in Table 4.3.

Hybrid aeration
Vertical shaft mechanical surface aerators (probably the more common) produce a mixture of water droplets and air bubbles which range from very fine to very coarse. The larger bubbles tend not to be stabilized by adsorbed detergent molecules. Mechanical surface aerators, therefore, are not α-sensitive. Fine-bubble diffused-air systems, by definition, produce fine bubbles which are stabilized by adsorbed

Table 4.3. Dissolved oxygen control methods

Method	Advantages	Disadvantages
Intermittent operation	Very cheap	Poor control. Risk of sludge settling in aeration tank and of poor quality effluent. Increased maintenance
Double wound motors	Cheap for small sizes	Poor control over limited range
Variable speed motors	Good control	Expensive if motors are large. Improvement in efficiency limited by losses in VS drive
Variable weir	Cheap	Very poor control. Only feasible for short retention time systems
Movement of aerator and gearbox	Good control, easily installed	Relatively high cost for small units
Movement of aerator relative to gearbox	Good control	Mechanically complex, resulting in high costs

detergent molecules. This means that they are less efficient when appreciable detergent concentrations are present. The hybrid system of aeration (Fig. 4.15) uses the mechanical surface aerators to degrade a significant proportion of the low α value pollutants in the initial phase of the process (Thomas *et al.*, 1989). The greater efficiency of the FBDA systems can then be used without significant α penalties. The relative proportions of the aeration tank which are aerated by the

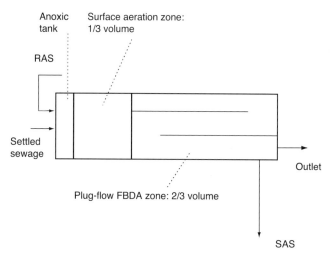

Fig. 4.15. A schematic diagram of a hybrid aeration plant

mechanical surface aerators and the FBDA systems will depend on the characteristics of the sewage and the tank geometry. Plants using hybrid aeration have been reported to produce good quality effluents consistently, with aeration efficiencies varying between 1·0 and 1·4 kg/kWh (Chambers et al., 1998).

Sludge settleability

The settling characteristics of a sludge are measured as the sludge volume index (SVI). The original method, which measured the volume occupied by the sludge after a quiescent period of settlement for 30 minutes in a 1 litre measuring cylinder, has now largely been discarded, although it is still useful as a day-by-day assessment of the settling characteristics of the sludge at one specific site. In some parts of the world, the diluted sludge volume index (DSVI) is now preferred. This involves settling 1 litre of MLSS and a series of diluted samples of MLSS. The sample used to determine the DSVI is that which gives a settled volume after 30 minutes of less than 25%. In the UK, the settleability of the sludge is usually measured as the stirred specific volume index (SSVI) (White, 1975). Its measurement is done on two samples, each of about 3 litres. Normally, MLSS and return activated sludge (RAS) are used. As with all the methods, the index is calculated from

$$\text{Index (SVI, DSVI or SSVI)} = \frac{\text{sludge volume after 30 min (\%)}}{\text{suspended solids concentration (\%)}}$$

(4.15)

For the SSVI, the value is quoted for a standard solids' concentration of 3·5 g/litre which is obtained by extrapolation or interpolation. A sludge with good settlement has a value of less than 90; one with an SSVI of more than 120 is usually considered to have 'bulked'.

Settleability of activated sludge is crucial to the success of the process. Successful design must include all the factors known to improve settleability and these are

(a) *Avoidance of septicity of sewage feed.* If the sewage arriving at the works is septic, pre-aeration may be required. Primary tanks should not be made so large that the duration of hydraulic retention leads to anaerobicity. It is sensible, always, to have at least two tanks, to ensure that only one can be used prior to design loading rates being reached. Iron may be used to precipitate insoluble sulphides if the sewage does becomes septic.

(b) *Adequate supply of dissolved oxygen throughout the aeration tanks.* It is essential to match the supply rate of oxygen to the demand rate. The minimum DO in a non-nitrifying plant would be 0·5 mg/litre and for a nitrifying plant it would be about 2·0 mg/litre.

(c) *Plug-flow configuration of aeration tanks.* Improved settleability is achieved when the floc-loading rate is high. As an approximate guide, the length to width ratio should exceed 12:1 with baffles to prevent back-mixing. Inlets for sewage feed and recycled sludge should be near to each other (or in a pre-mixing tank) and the inlet(s) should be at right angles to the length of the aeration tank if possible.

(d) *An adequate supply of essential nutrients (N, P and Fe).* The ratio of BOD:N:P should be 100:5:1 and soluble iron should be in excess of 0·1 mg/litre. Facilities for the addition of chemicals may be needed with some industrial wastewaters (e.g. brewery, pharmaceutical,

Fig. 4.16. An anoxic zone at the inlet of an FBDA-activated sludge plant

distillery wastes), but are also likely to be needed for domestic sewage when significant quantities of trade wastes are present or the sewage is septic.

(e) *An anoxic zone at the inlet end of a plug-flow nitrifying plant.* An anoxic zone is a tank into which recycled nitrifying activated sludge and sewage are fed with a retention period of about 0·7 h (based on average flow rate of sewage) (Fig. 4.16). It must not be aerated, either by back-mixing from the aeration tank downstream or by liquors cascading into the tank. It needs about 12 watts of mixing energy per m^3 to keep activated sludge suspended.

All nitrifying activated sludge plants should include an anoxic zone as it will achieve the following benefits and has no known disadvantages

- improved sludge settleability;
- reduced energy needed for treatment;
- reduced risk of de-nitrification in final settlement tanks;
- low nitrate content of final effluent;
- reduced growth of slimes on fine-bubble aeration diffusers down-stream;
- increased alkalinity (high pH value) aids nitrification.

Design of final settlement tanks

In most cases, radial flow tanks are used as activated sludge final settlement tanks. Essentially, their layout is similar to that used in primary settlement (Chapter 2) and although a floor slope of 10–12° is common (Fig. 4.17(a)) steeper slopes are not unknown (Fig. 4.17(b)).

The functions of these settling tanks are twofold: to separate activated sludge from final effluent and to produce a well-clarified effluent. Traditionally, the design of such tanks has relied on values of parameters, such as surface-loading rate (or up-flow velocity) and nominal retention time, selected by experience gained from previous design, operation and performance. Surface-loading rate will have an influence on the clarification of final effluent and it should not exceed the normally accepted value of 36 m^3/m^2 per day (1·5 m/h) at the maximum flow-rate of sewage (Anon, 1997). In some cases an even lower up-flow velocity of 1·2 m/h may be specified. Ekama and Marais (1986) have shown that the up-flow velocity at maximum flow should be varied with the settlement characteristics based on the DSVI. Thus, with an MLSS of 3·5 g/litre, if the DSVI is 100, a

(a)

(b)

Fig. 4.17. Final settlement tanks: (a) with 10° floor; (b) with 20° floor

maximum up-flow velocity of 1·8 m/h is acceptable. This needs to be reduced to 1·0 m/h if the DSVI deteriorates to 150 and then to 0·6 m/h if the DSVI increases to 200. Retention period is not so critical except that, for a given surface area, it will be directly proportional to the water depth and will influence the effective distribution of activated sludge onto the bottom of the tank. In practice, the design retention period is normally quoted as two hours (based on maximum flow-rate of sewage), and this corresponds with an average tank depth of 3 m recommended by the WRc and the CIRIA in Technical Report No. TR144 *The User Manual and Description for STOM* (Anon., 1981).

Mass-flux theory can be used to calculate the solids-loading rate from SSVI, the flow-rate of recycled sludge and the horizontal surface-area of a settling tank (White, 1975). The equation derived from this theory, which has been fully verified by full-scale design, operation and performance, gives the limiting flux (F_L) as

$$F_L = 307(S)^{-0.77}(Q_u/A)^{-0.68} \tag{4.16}$$

where F_L = limiting flux (kg/m^3 per hour), S = stirred specific volume index at 3·5 g/litre (ml/g), Q_u = under-flow (or recycle-rate) of returned sludge (m^3/h), A = horizontal-surface area of the settling tank (m^2). This limiting flux may not be exceeded. Thus

$$F_A \leq F \tag{4.17}$$

$$F_A = M(Q_u + Q_o)/A \tag{4.18}$$

where M = mixed liquor suspended solids (g/litre), Q_o = flow rate of sewage into the aeration tank (m^3/h), F_a = applied flux (kg/m^3 per hour). These equations have been expressed as a nomograph (Fig. 4.18) and can be used by the operator to avoid overloading the settling tank. Thus, a straight line running from the value of the SSVI to that of Q_u/A can be extrapolated to show the limiting flux. Similarly, a line from the MLSS concentration to the value of $(Q_o + Q_u)/A$ can be extended to give the applied flux. If F_A is greater than F_L, the nomograph can be used to determine how the plant operation should be modified.

For design purposes, equations (4.16), (4.17) and (4.18) can be rearranged to give an equation to calculate the total plan area of the settlement tanks

$$A = \frac{M^{3.125}S^{2.41}[Q_u^{0.32} + (Q_o/Q_u^{0.68})]^{3.125}}{307^{3.125}} \tag{4.19}$$

This can be simplified to

$$A_A = \frac{M^3 Q}{2000f} \tag{4.20}$$

where A_A = **approximate** surface area (m^2), Q = average flow rate of sewage which is equal to the flow rate of recycled sludge (m^3/d), f = factor which varies with SSVI as indicated below.

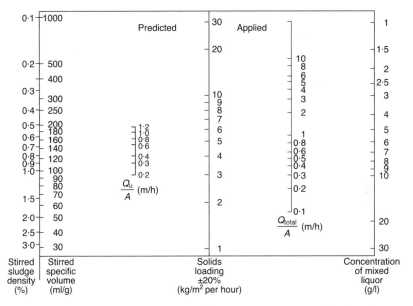

Fig. 4.18. *Nomograph for the calculation of limiting and applied loading* (White, 1975)

SSVI	f
75	2·1
100	1·0
125	0·6
150	0·4

Note that all the above equations can **only** be applied when Q_u/A is less than 1·2 m/h and more than 0·2 m/h. Also, the upward-flow velocity Q_o/A must not exceed 1·5 m/h.

The benefit of producing a sludge of low SSVI is also clear, as the cross-sectional area will be significantly reduced. For example, the area required at an SSVI of 75 ml/g is only 0·2 of that for 150 ml/g.

Aeration tank configuration

In terms of hydraulic mixing, two basic plant configurations can be defined: plug-flow and completely mixed (Fig. 4.19(a) and (b)). In

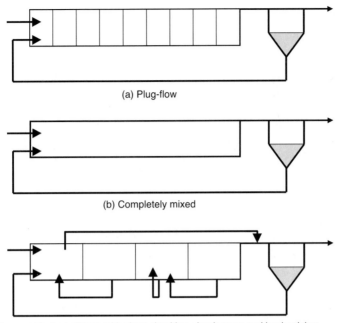

(a) Plug-flow

(b) Completely mixed

(c) Nearer actual conditions with short-circuiting, dead zones and back mixing

Fig. 4.19. *Mixing regimes in an activated sludge aeration tank*

practice, pure plug-flow conditions will not exist. There will always be a degree of back-mixing and short-circuiting (Fig. 4.19(c)). Therefore, the actual conditions will need to be found by tracer studies. Tracer studies are used to determine the degree of longitudinal mixing within an aeration tank. Typically, lithium chloride or rhodamine WT would be used. After dosing, samples would be taken from the outlet of the aeration tank and the concentration of the tracer (C) would be measured at fixed time intervals (t). The degree of longitudinal mixing in an aeration tank is quantified in terms of the Dispersion Number as described by Levenspiel (1972). This has values between 0, for plug flow tanks, and ∞ for completely mixed tanks. In the analysis of tracer study data, it is useful to work with dimensionless variables (normalized) which measure time in relation to the mean residence time (t_R) and tracer concentration in relation to the initial tracer concentration (C_O).

Thus, normalized time, θ, is given by t/t_R and the normalized concentration, E, by C/C_O. Plots of E against θ give curves of the type shown in Fig. 4.20. The data are used to calculate the variance, σ^2, the dispersion number and the number of tanks in series. This is an

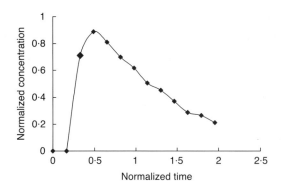

Fig. 4.20. A typical exit age distribution curve

alternative way of quantifying the degree of longitudinal mixing and is the number of hypothetical completely mixed tanks which, if operated in series, have the same mixing characteristics as the tank being tested. Thus, a completely mixed tank is classified as one tank in series and a perfect plug-flow tank is equivalent to an infinite number of tanks in series.

Variance (σ^2) is given by

$$\sigma^2 = \frac{\Sigma C \theta^2}{\Sigma C} - 1 \tag{4.21}$$

The dispersion number, D/uL, is given by

$$\sigma^2 = 2(D/uL) - 2(D/uL)^2(1 - e^{-uL/D}) \tag{4.22}$$

where D = dispersion coefficient, u = average fluid velocity, L = total aeration tank length. Equation (4.22) is solved by iteration starting with $\sigma^2 = 2(D/uL)$. The number of tanks in series (N) can be derived from

$$\sigma^2 = \frac{1}{N} \tag{4.23}$$

For FBDA systems, the degree of mixing can be estimated, to within 15%, from (Chambers and Jones, 1985)

$$N = \frac{7 \cdot 4}{WH} LQ(1 + r) \tag{4.24}$$

where L = aeration tank length (28 to 500 m), W = aeration tank width (2 to 20 m), H = aeration tank depth (2·4 to 6 m), Q = average sewage flow-rate (m^3/s), r = recycle ratio.

Activated sludge process

Sludge production

The amount of sludge produced will depend on the quantity of inert suspended solids in the incoming feed, the sludge-loading rate (Fig. 4.6), and, to a lesser extent, the temperature of the mixed liquor. Typically, Y will be 0·7–1·0 kg dry solids per kg BOD removed. For settled domestic sewage, the sludge yield (Y) in kg dry weight of solids/kg BOD removed can be derived from Fig. 4.6 or from empirical relationships such as

$$Y = X + 0\cdot 68 \left(\frac{(B_s - B_e)}{Mt_r} \right)^{-0\cdot 24} \qquad (4.25)$$

where B_s = BOD of sewage (mg/litre), B_e = BOD of effluent (mg/litre), M = MLSS (mg/litre), t_r = aeration period of sewage (days), X = inert component coming from settled sewage (kg dry solids/kg BOD removed). The value selected for the inert debris (X) may vary, depending on the industrial content of the sewage. Therefore, wherever possible, it should be verified by measurements and calculation from existing data. If 'p' is the percentage of inert material in the solids in the settled sewage

$$X = \frac{Q \times \text{inert SS}}{Q \times \Delta \text{BOD}} = \frac{Q \times \text{TSS} \times p}{Q \times \Delta \text{BOD}} = \left(\frac{\text{TSS}}{\Delta \text{BOD}} \right) p \qquad (4.26)$$

Thus, having works-specific data for the total suspended solids (TSS) and BOD, X can be derived from plots such as those shown in Fig. 4.21.

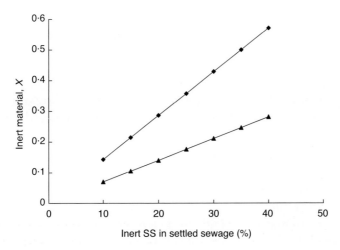

Fig. 4.21. *The effect of varying inert SS on X for Works A (◆) and Works B (▲)*

These equations can provide a guide to the weight of dry solids likely to be produced and, therefore, the amount of sludge which must be wasted. They will not give any information about the volume of the sludge nor its wet weight.

For a non-nitrifying plant, operated at a relatively-high sludge-loading rate, the surplus activated sludge is likely to be derived from recycled sludge containing about 4 g/litre solids (0·4% w/v solids). A nitrifying plant, operated with a long sludge age, low sludge-loading rate and higher MLSS, is likely to have a surplus sludge with a solids content of 8 to 10 g/litre (0·8 to 1% w/v solids) so that only half the volume of sludge would be produced compared with the non-nitrifying plant. In addition, activated sludge from a nitrifying plant may not be further consolidated without the addition of large quantities of chemicals, which may not be cost-effective.

Sludge wastage rate

Wastage of sludge is essential to prevent a continual increase of MLSS, and it will affect the sludge age. However, there should not normally be a need to vary the volumetric rate frequently (maybe four or five times yearly would be sufficient, unless the recycle rate was varied).

Reasons for varying the wastage rate include

- the MLSS content is too high and the solids loading rate of the settlement tank is above the predicted value (based on SSVI, MLSS and flow-rates of sewage and recycled sludge);
- the need to reduce sludge age to prevent nitrification or the need to increase sludge age to achieve nitrification (changes in mixed liquor temperature — summer and winter — may result in a need to vary the MLSS);
- the BOD load into the works has increased and, therefore, the MLSS has to be increased to keep sludge loading constant;
- substances toxic to bio-oxidation have been discharged into the sewage so that an increase in sludge age is required to maintain the effluent quality with a reduced growth rate of bacteria.

Sludge recycle rate

A constant rate at the minimum design value should be the normal operational mode. With a nitrifying plant, the rate may have to be higher to achieve a significant proportion of de-nitrification in the anoxic zone (but 1·5 times the average flow rate of sewage would generally be the maximum).

The minimum recycle rate should be $0\cdot2\,m^3$ per m^2 horizontal-surface area of the settlement tank per hour and the maximum $1\cdot2\,m^3/m^2$ per hour. If the rate is normally kept at the minimum, then it can be increased to prevent the applied solids loading rate exceeding the predicted rate. The lower the recycle rate, the greater will be its solids content and hence the smaller will be the volumetric wastage rate. Thus, when the rate of recycle is varied, the wastage rate must be changed to maintain the same 'dry weight' wastage of sludge.

Operational problems

There are basically two problems which affect activated sludge plants, i.e. bulking and the production of stable foam. Both are associated with dominant populations of filamentous microbes in the sludge.

Filamentous species

In looking at these problems, there are two questions which need to be addressed: do all filaments have the same effect and what causes the filaments to become dominant? However, before addressing these questions, it is necessary to know how to identify the filamentous species in activated sludge. The early reports about bulking linked the filamentous species *Sphaerotilus natans* to the problem (Morgan and Beck, 1928) and there was a general belief, until the 1970s, that this was the only filamentous bacteria likely to cause bulking. This was incorrect, as was shown by Eikelboom (1975) who identified 26 types of filamentous micro-organisms in activated sludge. It must be stressed that these were types and were not a strict taxonomic identification. Indeed, some types were identified only by a numerical code. Nevertheless, Eikelboom's work provided a method by which sewage works operators could classify the filaments in their sludge and set about categorizing the behaviour of sludges dominated by the different types.

The identification of filamentous species is achieved by the Eikelboom technique (Eikelboom and van Buijsen, 1983; Jenkins *et al.*, 1993). This uses microscopic examinations of wet mount samples, Gram stained samples and Neisser stained samples in conjunction with a simple 'key' (Fig. 4.22). The abundance of the filaments is assessed subjectively using the scale given in Table 4.4 (Jenkins *et al.*, 1993).

Wastewater treatment and technology

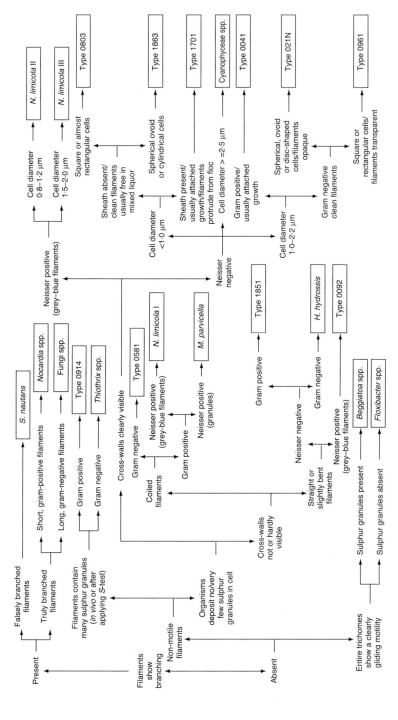

Fig. 4.22. Identification key for filamentous bacteria in activated sludge (Eikelboom, 1982)

Table 4.4. Subjective scoring of filament abundance (Jenkins et al., 1993)

Score	Abundance	Description
0	None	
1	Few	Filaments present but only in the occasional floc
2	Some	Filaments commonly observed but not present in all flocs
3	Common	Filaments observed in all flocs but only 1–5 per floc
4	Very common	Filaments observed in all flocs at a density of about 5–10 per floc
5	Abundant	Filaments observed in all flocs at a density of about >20 per floc
6	Excessive	Filaments observed in all flocs with the appearance of more filaments than flocs

In recent years, phylogenetic analyses, based on 16S rDNA sequencing, has been used increasingly to clarify filament taxonomy. As a result, *Nocardia amarae* (Fig. 4.23) has been reclassified as *Gordona amarae* (Ruimy et al., 1994), 021N (Fig. 4.24) as *Thiothrix eikelboomii* (Howarth et al., 1999) and *M. parvicella* (Fig. 4.25) has been characterized as an actinomycete (Blackall et al., 1994). This technique has also shown that the morphological classification is not always accurate. For example, five isolates which were identified morphologically as Eikelboom type 1863 were shown by genetic analysis to be polyphyletic (Seviour et al., 1997).

A series of national surveys provides a partial answer to the question about all filaments having the same effect. The results (Table 4.5) show the ranking of the occurrence of filamentous species in bulking sludges and demonstrate that some types are more likely to promote bulking than others.

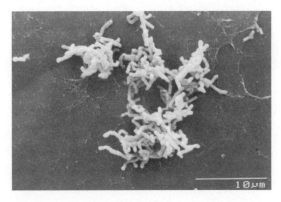

Fig. 4.23. Nocardia amarae-*like organism*

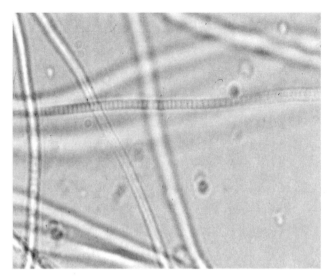

Fig. 4.24. 021N magnified ×1000 (courtesy Dr M. Kay, PIRA International)

For some time the precise growth conditions or growth requirements for specific filamentous species were limited due to the problems of isolating them in pure culture. However, this situation is gradually being remedied (Seviour and Blackall, 1999). The different environmental conditions which have been associated with the different filaments are shown in Table 4.6. These provide indications about possible means for controlling their dominance.

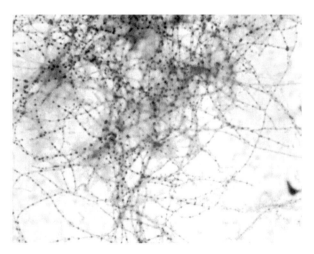

Fig. 4.25. Microthrix parvicella

Table 4.5. A ranking of filamentous microbes in bulking activated sludge

Ranking	Species		
	Germany (Wagner, 1982)	The Netherlands (Eikelboom, 1977)	USA (Jenkins et al., 1984)
1	021N	M. parvicella	Nocardia spp.
2	M. parvicella	021N	1701
3	0041	H. hydrossis	021N
4	S. natans	0092	0041
5	Nocardia spp.	1701	Thiothrix spp.
6	H. hydrossis	0041	S. natans
7	N. limicola	S. natans	M. parvicella
8	1701	0581	0092

Bulking

As the total filament length increases, so does the SSVI (Fig. 4.26) (Sezgin et al., 1980). When the SSVI is greater than, nominally, 120, the top of the sludge blanket in the final settlement tank becomes close to top water level, and there is a significant risk of losing solids in the final effluent and exceeding the consent concentrations. Lee et al. (1983) examined the relationships between all three settlement indices and the total extended filament length (TEFL), and found that the DSVI showed the best correlation, with the DSVI increasing sharply above 150 when the TEFL exceeded 30 km/g.

A sequence of procedures for dealing with a bulking sludge has been formulated by the Water Research Centre (Fig. 4.27) (Tomlinson and

Table 4.6. Environmental conditions for the proliferation of filamentous bacteria

Conditions	Likely filaments
Low loading; extended aeration plants	Winter...M. parvicella, Summer...0092
Conventional loading; full/partial nitrification	021N, M. parvicella, 1701, 0803, H. hydrossis
Conventional loading; no nitrification	021N, 1701, H. hydrossis, S. natans
Low DO	1701, S. natans, H. hydrossis
Septicity	Thiothrix, Beggiatoa, 021N
Nutrient imbalance	Thiothrix, S. natans, 021N
Low F:M	M. parvicella, H. hydrossis, 021N, 0041, Nocardia, 0675, 0092, 0961, 0803

Wastewater treatment and technology

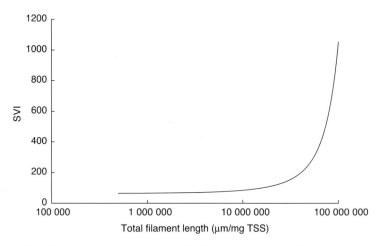

Fig. 4.26. The relationship between total filament length and sludge settlement

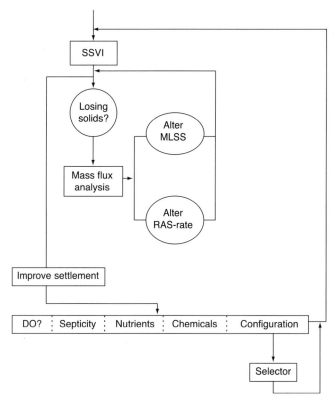

Fig. 4.27. Control sequence for bulking sludges

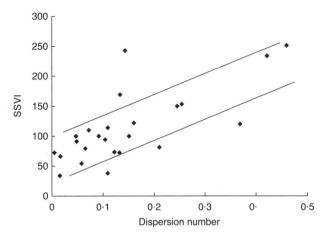

Fig. 4.28. The relationship between activated sludge settlement characteristics and the Dispersion number (Tomlinson and Chambers, 1979)

Chambers, 1984). As a first step, a mass flux analysis of the final settlement tanks should be carried out to determine whether these tanks are being overloaded. The next stage is to examine the mixing characteristics of the aeration tank because it has been shown that, as the degree of plug-flow increases, the settlement characteristics of the sludge improve (Fig. 4.28) (Tomlinson and Chambers, 1979). It is not always possible to make significant changes to the mixing regime of an existing plant, but where this is possible it can have a positive effect as was shown by Humphries (1982). The flow regimes before and after modification are shown in Fig. 4.29. The SSVI for the sludge was initially 150–200. Two months after converting the flow it had fallen to less than 100. Other factors which can be checked are the nutrient balance, the DO concentration and whether there is any septicity, in either the incoming sewage or the return sludge. Finally, consideration should be given to the use of a selector.

A selector is a small volume tank located immediately before the aeration tank in which the RAS and the incoming settled sewage are contacted giving a high, instantaneous food to mass (F:M) ratio. It is a device which discriminates against filaments and favours floc-formers on the basis of their relative kinetic properties. Depending on the oxic-regime which prevails, the selector may also achieve additional treatment.

Aerobic selectors have controlled the low F:M filaments (0092, 0914, 1851, *M. parvicella*, 0675 and 0041) when the system was

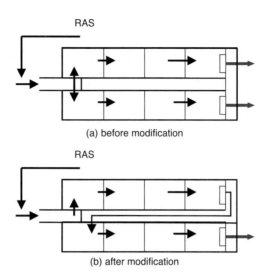

Fig. 4.29. Flow pattern modifications to improve longitudinal mixing (Humphries, 1982)

totally aerobic. However, when there was intermittent aeration, control was inadequate (Gabb et al., 1991).

There are no sound design criteria, despite the selector concept having been around for some time. There ought to be a design route through the kinetic constants (μ_{max} and K_s, defined after equation (1.6)), but much of the data which are available in the literature for pure cultures of filamentous bacteria are for specific chemicals, such as acetate, and, as such, are not of much use. Hence, perceived wisdom must be used (Wanner, 1998) and that is

- a hydraulic retention time (based on Q and Q_u) of 10–20 minutes;
- a good substrate gradient across the tank;
- plug-flow wherever possible;
- a floc loading of about 3 kg BOD/kg MLSS per day.

Anoxic selectors will act against 021N, which is one of the main causes of bulking in the UK, *N. limicola II* and *S. natans*. They have also been found to be able to suppress *Nocardia* but not to control *M. parvicella*. Anoxic selectors will achieve de-nitrification when nitrate is present in the RAS. The process of de-nitrification will also remove BOD/COD. Thus, it reduces the load on the aeration tank aerators and, therefore, reduces the costs of aeration. The design is based on a hydraulic retention time of 10–30 minutes and a loading of 6 kg BOD/kg per day (Wanner, 1998).

Fig. 4.30. Foam covering a final settlement tank

The choice of regime depends on the filamentous species requiring control, and knowledge is not always perfect. Indeed, it has been said (Chudoba and Pujol, 1994) that the diversity of the physiology of filaments means that there is no universal method to control the growth of all filamentous species. The choice of regime is clouded by the different modes of operation in the main treatment system.

Foam formation
Stable foams are a relatively new phenomenon (Anon, 1969; Soddell and Seviour, 1990). They occur world-wide and are a dark brown, greasy scum which usually forms on the surface of aeration tanks (Fig. 4.30). They become a serious problem when they move to the settlement tank and are discharged with the final effluent. They are also a problem if they over-flow from sumps or wet-wells (Fig. 4.31). The foams are viscous (more than 2500 times water), hydrophobic and stable (half-life = 40–60 hours). Foam is formed when the sludge becomes dominated by filaments. As can be seen from Table 4.7 the most problematical species are *Nocardia* spp. and M. *parvicella*. In the UK, M. *parvicella* is the most common foam former.

Obviously, some of the points which were discussed in the earlier section on bulking are also applicable to foaming filaments. In addition, there are some criteria, which, whilst not proven scientifically, are firmly

Fig. 4.31. Cleaning up a foam excursion

embedded in the folk-lore about foam. One specific point is that the presence of 'fats and oils' promotes foaming. Certainly, the hydrophobic filamentous bacteria have a metabolic advantage when fats and oils form a significant proportion of the incoming COD, but the concept is yet to be proved.

No-one knows why a plant will start to foam, so attention must focus on control. The options are (Tipping, 1995; Wanner, 1998)

(a) *Lower the sludge age.* The filaments are slow growing. Therefore, a reduction in the sludge age will selectively favour the quicker-growing floc-formers (Richards et al., 1990), but care must be taken with

Table 4.7. A ranking of filamentous microbes in bulking activated sludge

Species	France	South Africa
M. parvicella	1	1=
0675	2	
Nocardia	3	3
0092	4	1=
0041		4
0803		5

Activated sludge process

nitrifiers if the plant has an ammonia consent. The critical value for M. parvicella elimination is reported as being 5 days (Jenkins, 1992).
(b) *Use anti-foams.* The data on the success rate of using anti-foams tend to suggest that their effectiveness is plant specific. They can reduce the extent of the problem but will not eliminate it. They are expensive and they will impair oxygen transfer.
(c) *Use biocides.* Chlorination is a popular technique in North America (Jenkins et al., 1993) and one which can be effective against filaments. The recommended dose is 2–10 kg/t MLSS for a non-nitrifying sludge and less than 5 kg/t MLSS when nitrification is required (Chudoba et al., 1985). If chlorination is used, the question about dispersing unknown chlorination by-products into the environment has to be considered. Ozone has been assessed as a means of controlling bulking and has been found to be successful. In addition, ozone treatment improves the removal of organics and aids nitrification (van Leeuwen and Pretorius, 1988). However, it is an expensive means of treatment.
(d) *Water sprays.* Spraying with water is really no more than a 'stopgap' measure. It does not remove the problem, it merely minimizes the risk of exceeding consent conditions. Heavy spraying will keep a final tank clear of surface foam once the foam has formed. Sprays permanently fitted to the bridges of final tanks can prevent the onset of foam coverage (Fig. 4.32).
(e) *Selectors.* Again the choice of regime depends on the filamentous species requiring control.

M. parvicella is a species which causes bulking and foaming. However, it appears to do one or the other. Indeed, an examination of three activated sludge plants showed that, although there was an identical dominance of the sludges by M. parvicella, one was bulking, one was producing foam and the other was operating without either of these problems. A possible explanation of this is that the surface characteristics were different (Forster, 1996) (Fig. 4.33). This concept is supported by recent work which suggests that any hydrophobic bacterial (rods, cocci as well as filaments) can cause foaming, with the hydrophobicity being the important feature rather than the morphology (Davenport et al., 2003).

Mathematical models
Although the activated sludge process has been in use for some 80 years, the application of mathematical modelling only commenced in the

Wastewater treatment and technology

Fig. 4.32. Water sprays on a final settlement tank

1970s. Most main-stream models are based on Monod kinetics and the models which are now available have evolved almost in line with the development of computers. Initially, the process modelling focused largely on the behaviour of the aeration tanks for the removal of carbonaceous and nitrogenous pollutants, but, in 1991, a dynamic model for the settling tank became available (Tackacs et al., 1991), thus enabling the complete activated sludge process to be modelled. One of the key models is that developed by the International

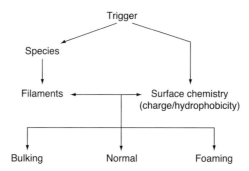

Fig. 4.33. Integrated interactions to explain bulking and foaming

106

Association of Water Quality (IAWQ) (Henze et al., 1987, 1995). This is based on COD and, therefore, for the UK, where most of the sewage works monitoring is based on BOD, it is perhaps not the most convenient model. Furthermore, the organic components must be divided into biodegradable and non-biodegradable COD-fractions and then further subdivided into soluble and particulate material. This requires a considerable analytical input which is not always justifiable. In this context, it is worth noting that there has recently been a detailed examination of the characteristics of wastewater (Rössle and Pretorius, 2001) which may be of use, in conjunction with the default value in the models, in guiding the uninitiated. The STOAT model which has been developed by the Water Research Centre (Smith and Dudley, 1998) is a dynamic model. Its analysis of the activated sludge process can be made either with the COD-based IAWQ models or the BOD-based models developed by Jones (1978). While the literature contains a number of examples of the IAWQ models and their application (e.g. Finnson, 1993; Kappler and Gujer, 1994), work based on the Jones model is less frequently reported. Important issues in the use of any model are how well it performs, how easily it can be calibrated and which default values need to be changed. It must be remembered that, whatever type of model is used, the validity/applicability of the constants is probably the weakest link.

Sequence for design

There are two options for designing an activated sludge plant. Both of them necessitate making assumptions for certain parameters — what would be called design specifications. These are shown in Table 4.8.

Table 4.8. Parameters to be specified for design

Parameter	Route 1	Route 2
Q_o/A		✓
RAS concentration	✓	✓
SSVI	✓	✓
OLR		✓
θ	✓	
Y	✓	
MLSS concentration	✓	

Route 1
With this approach the aeration tank volume (V in m^3) would be derived independently from the plan area of the settlement tanks (A in m^2). Thus

$$V = \frac{L \times \theta \times Y}{M} \tag{4.27}$$

$$A = \frac{M^{3 \cdot 125} S^{2 \cdot 41} [Q_u^{0 \cdot 32} + (Q_o/Q_u^{0 \cdot 68})]^{3 \cdot 125}}{307^{3 \cdot 125}} \quad \text{see equation (4.19)}$$

where L = BOD load (kg/d), θ = sludge age (days), Y = sludge yield (kg SS/kg BOD applied), M = MLSS concentration (kg/m^3), Q_o = average settlement tank overflow rate (m^3/h), Q_u = average settlement tank underflow rate (m^3/h).

Route 2
With this approach a set of integrated calculations are used to derive the values of A and V. Thus, select A to be compatible with up-flow velocity requirements

$$\frac{FTFT}{A} = 45 \, \text{m}^3/\text{m}^2 \text{ per day} \tag{4.28}$$

Specify the design SSVI and calculate the design MLSS

$$A = \frac{M^{3 \cdot 125} S^{2 \cdot 41} [Q_u^{0 \cdot 32} + (Q_o/Q_u^{0 \cdot 68})]^{3 \cdot 125}}{307^{3 \cdot 125}} \quad \text{see equation (4.19)}$$

Calculate the tank volume from the OLR and the MLSS

$$OLR = \frac{BQ}{MV} \quad \text{see equation (4.1)}$$

In both cases, Q_u would be determined from a mass balance across the settlement tank.

References
Anon. (1969) Milwaukee mystery: unusual operating problems develop, *Wat. Sew. Wks.*, **116**, 213.
Anon. (1981) Sewage treatment optimization model, user manual and description, *Technical Report, TR 144*, Water Research Centre, Medmenham.
Anon. (1997) *Handbooks of UK Wastewater Practice: Activated Sludge Treatment*, CIWEM.

Badger, R. B., Robinson, D. D. and Kiff, R. J. (1975) Aeration plant design: Derivation of basic data and comparative performance studies, *Wat. Pollut. Control*, **74**, 415–29.

Blackall, L. L., Seviour, E. M., Cunningham, M. A., Seviour, R. J. and Hugenholtz, P. (1994) *Microthrix parvicella* is a novel, deep branching member of the actinomycetes subphylum, *Syst. Appl. Microbiol.*, **17**, 513–18.

Chambers, B. and Jones, G. L. (1985) Energy saving by fine-bubble aeration, *Wat. Pollut. Control*, **84**, 70–86.

Chambers, B., Upton, J. and Greenhaigh, S. H. (1998) The development and use of hybrid aeration proceses, *J. CIWEM*, **12**, 322–9.

Chudoba, J., Cech, J. S., Farkac, J. and Grau, P. (1985) Control of activated sludge filamentous bulking — experimental verification of a kinetic selection theory, *Water Res.*, **19**, 191–6.

Chudoba, P. and Pujol, R. (1994) Kinetic selection of microorganisms by means of a selector — twenty years of progress: history, practice and problems, *Wat. Sci. Technol.*, **29**, 177–80.

Churchley, J., Thomas, A. and Stokes, L. (2000) A practical approach to process problems, *Conf. Activated Sludge: The Lessons of the Last Decade*, Nottingham, March, Chartered Institution of Water and Environmental Management.

Curds, C. R. (1966) An ecological study of ciliated protozoa in activated sludge, *Oikis*, **15**, 282–9.

Curds, C. R., Cockburn, A. and Vandyke, J. M. (1968) An experimental study of the role of ciliated protozoa in the activated sludge process, *Wat. Pollut. Control*, **67**, 312–29.

Davenport, R., Pickering, R. and Curtis, T. P. (2003) Activated sludge foaming, gaining insight with molecular tools, *Paper presented at CIWEM Seminar Foaming and Bulking of Activated Sludge — Problems and Solutions*, May, Wakefield.

Downing, A. L. (1968) Factors to be considered in the design of activated sludge plant, in *Advances in Water Quality Improvement*, Gloyna, E. F. and Eckenfelder, W. W. (eds), Water Resources Symposium No. 1, University of Texas Press, pp 190–202.

Eckenfelder, W. W. and O'Connor, D. J. (1961) *Biological Waste Treatment*, Pergamon Press, London.

Eikelboom, D. H. (1975) Filamentous organisms observed in activated sludge, *Water Res.*, **9**, 365–88.

Eikelboom, D. H. (1977) Identification of filamentous organisms in bulking activated sludge, *Prog. Wat. Technol.*, **8**, 153–62

Eikelboom, D. H. (1982), Microscopic sludge investigation in relation to treatment plant operation, in *Bulking of Activated Sludge*, Chambers, B. and Tomlinson, E. J. (eds), Ellis Horwood Ltd, Chichester, pp 47–60.

Eikelboom, D. H. and van Buijsen, M. S. S. (1983) *Microscopic Sludge Investigation Manual*, TNO Research Institute, Delft.

Ekama, G. A. and Marais, G. v. R. (1986) Sludge settleability and secondary settling tank design procedures, *Wat. Pollut. Control*, **85**, 101–13.

Finnson, S. (1993), Simulation of a strategy to start up nitrification at Bromma sewage plant using a model based on the IAWQPRC model No 1, *Wat. Sci. Technol.*, **28**, 185–96.

Forster, C. F. (1996) Aspects of the behaviour of filamentous microbes in activated sludge, *J. CIWEM*, **10**, 290–4.

Forster, C. F. and Dallas-Newton, J. (1980) Activated sludge settlement — some suppositions and suggestions, *Wat. Pollut. Control*, **79**, 338–51.

Gabb, D. M. D., Still, D. A., Ekama, G. A., Jenkins, D. and Marais, G. v. R. (1991) Selector effect on filamentous bulking in long sludge age activated sludge systems, *Wat. Sci. Technol.*, **23(4–6)**, 867–77.

Henze, M., Grady, C., Gujer, W., Marais, G. and Matsuo, T. (1987) *Activated Sludge Model No. 1*, IAWPRC Task Group on Mathematical Modelling for Design and Operation of Biological Wastewater Treatment, IAWQ, London, UK.

Henze, M., Gujer, W., Marais, G., Matsuo, T., Mino, M. and Wentzel, M. (1995) *Activated Sludge Model No. 2*, IAWQ Task Group on Mathematical Modelling for Design and Operation of Biological Wastewater Treatment Processes, IAWQ, London, UK.

Horan, N. J. and Eccles, C. R. (1986) The potential for expert systems in the operation and control of activated sludge plants, *Process Biochem.*, **21**, 81–5.

Howarth, R., Unz, R. F., Seviour, E. M., Seviour, R. J., Blackall, L. L., Pickup, R. W., Jones, J. G., Yaguchi, J. and Head, I. M. (1999) Phylogenic relationships of filamentous sulfur bacteria (*Thiothrix* spp. and *Eikelboom* type 021N bacteria) isolated from wastewater-treatment plants and description of *Thiothrix eikelboomii sp nov.*, *Thiothrix unzii sp nov.*, *Thiothrix fructosivorans sp nov.* and *Thiothrix defluvii sp nov.*, *Int. J. Sys. Bact.*, **49**, 1817–27.

Humphries, M. D. (1982) The effect of a change in longitudinal mixing on the settleability of activated sludge, in *Bulking of Activated Sludge*, Chambers, B. and Tomlinson, E. J. (eds), Ellis Horwood Ltd, Chichester, pp 261–4.

Jenkins, D. (1992) Towards a comprehensive model of activated sludge bulking and foaming. *Wat. Sci. Technol.*, **25**, 215–30.

Jenkins, D., Richard, M. D. and Neethling, J. B. (1984) Causes and control of activated sludge bulking, *Wat. Pollut. Control*, **83**, 455–72.

Jenkins, D., Richard, M. D. and Daigger, G. T. (1993) *Manual on the Causes and Control of Activated Sludge Bulking and Foaming*, Lewis Publishers, New York, USA.

Johnstone, D. W. M., Rachwal, A. J. and Hanbury, M. J. (1983) General aspects of the oxidation ditch process, in *Oxidation Ditches in Wastewater Treatment*, Barnes, D., Forster, C. F. and Johnstone, D. W. M. (eds), Pitman Publishing Ltd, London, pp 41–74.

Jones, G. L. (1978) A mathematical model for bacterial growth and substrate utilisation in the activated-sludge process, in *Mathematical Models in Water Pollution Control*, James, A. (ed), John Wiley and Sons, London, pp 265–79.

Kappeler, J. and Gujer, W. (1994) Development of a mathematical model for 'aerobic bulking', *Water Res.*, **28**, 303–10.

Lee, S. E., Koopman, B., Bode, H. and Jenkins, D. (1983) Evaluation of alternative sludge settleability indices, *Water Res.*, **17**, 1421–6.

Levenspiel, O. (1972) *Chemical Reaction Engineering*, John Wiley and Sons, New York.

Lister, A. R. and Boon, A. G. (1973) Aeration in deep tanks: An evaluation of a fine-bubble diffused-air system, *Wat. Pollut. Control*, **72**, 590–605.

Marais, G. v. R. (1973) The activated sludge process at long sludge ages, *Research Report No. W3*, Department of Civil Engineering, University of Cape Town.

Morgan, E. H. and Beck, A. J. (1928) Carbohydrate wastes stimulate growth of undesirable filamentous organisms in activated sludge, *Sew. Wks. J.*, **1**, 46–51.

Morgan, J. W. and Forster, C. F. (1992) A comparative study of the sonication of anaerobic and activated sludges, *J. Chem. Tech. Biotech.*, **55**, 53–8.

Pipes, W. O. (1967) Bulking of activated sludge, *Adv. Appl. Microbiol.*, **9**, 185–234.

Rees, J. T. and Skellett, C. F. (1974) Performance of a Passavant Mammoth aeration plant, *Wat. Pollut. Control*, **73**, 608–20.

Richards, T., Nungesser, P. and Jones, C. (1990) Solution of *Nocardia* foaming problems, *Res. J. Wat. Pollut. Control Fed.*, **62**, 915–19.

Rössle, W. H. and Pretorius, W. A. (2001) A review of characterisation requirements for in-line prefermenters: Wastewater characterisation, *Water SA*, **27**, 405–12.

Ruimy, R., Boiron, P., Biovin, V. and Christen, R. (1994) A phylogeny of the genus *Nocardia* deduced from the analysis of small-subunit DNA sequences, including the transfer of *Nocadia amarae* to the genus *Gordona* as *Gordona amarae* nov., *FEMS Micobiol. Lett.*, **123**, 261–8.

Sanin, F. D. and Vesilind, P. A. (2000) Bioflocculation of activated sludge: the role of calcium ions and extracellular polymers, *Environ. Technol.*, **21**, 1405–12.

Severn Trent Water (1995) Personal communication.

Seviour, E. M., Blackall, L. L., Christensson, C., Hugenholtz, P., Cunningham, M. A., Bradford, D., Stratton, H. M. and Seviour, R. J. (1997) The filamentous morphotype Eikelboom 1863 is not a single genetic entity, *J. Appl. Microbiol.*, **82**, 411–21.

Seviour, R. J. and Blackall, L. L. (1999) Current taxonomic status of filamentous bacteria found in activated sludge plants, in *The Microbiology of Activated Sludge*, Seviour, R. J. and Blackall, L. L. (eds), Kluwer Academic Publishers, The Netherlands, pp 122–46.

Sezgin, M., Jenkins, D. and Parker, D. S. (1978) A unified theory of filamentous activated sludge bulking, *J. Wat. Pollut. Control Fed.*, **50**, 362–81.

Sezgin, M., Jenkins, D. and Palm, J. C. (1980) Floc size, filament length and settling properties of prototype activated sludge plants, *Prog. Wat. Technol.*, **12(3)**, 171–82.

Smith, M. and Dudley, J. (1998) Dynamic process modelling of activated sludge plants, *J. CIWEM*, **12**, 346–56.

Soddell, J. A. and Seviour, R. J. (1990) Microbiology of foaming in activated sludge plants, *J. Appl. Bact.*, **69**, 145–76.

Steiner, A. E., McLaren, D. A. and Forster, C. F. (1976) The nature of activated sludge flocs, *Water Res.*, **10**, 25–30.

Tackacs, L., Patry, G. G. and Nolasco, D. (1991) A dynamic model of the clarification–thickening process, *Water Res.*, **25**, 1263–71.

Tewari, P. K. and Bewtra, J. K. (1982) Alpha and beta factors for domestic wastewater, *J. Wat. Pollut. Control Fed.*, **54**, 1281–7.

Thomas, V. K., Chambers, B. and Dunn, W. (1989) Optimisation of aeration efficiency: A design procedure for secondary treatment using a hybrid aeration system, *Wat. Sci. Technol.*, **21**, 1403–19.

Tipping, P. J. (1995) Foaming in activated sludge processes: An operator's overview, *J. CIWEM*, **9**, 281–9.

Tomlinson, E. J. and Chambers, B. (1979) The effect of longitudinal mixing on the settleability of activated sludge, *Technical Report TR122*, Water Research Centre, Medmenham.

Tomlinson, E. J. and Chambers, B. (1984) Control strategies for bulking sludge, *Wat. Sci. Technol.*, **16(10/11)**, 15–34.

van Leeuwen, J. and Pretorious, W. A. (1988) Sludge bulking control with ozone, *J. IWEM*, **2**, 223–7.

Wagner, F. (1982) Studies of the cause and prevention of sludge bulking in Germany, in *Bulking of Activated Sludge*, Chambers, B. and Tomlinson, E. J. (eds), Ellis Horwood Ltd, Chichester, pp 29–40.

Wanner, J. (1998) Stable foams and sludge bulking: The largest remaining problems, *J. CIWEM*, **12**, 368–74.

White, M. J. D. (1975) Settling of activated sludge, *Technical Report, TR 11*, Water Research Centre, Medmenham.

5
Bio-oxidation — other aspects

Introduction
Although the activated sludge process and trickling filters tend to dominate the market for the treatment of municipal wastewater, there are a number of other types of biological process which must be considered. Some of these have a very distinct niche and can offer advantages over the more traditional processes.

Biological contactors
These are fixed film processes and two distinct options can be considered: rotating biological contactors (RBCs) and submerged biological contactors (SBCs). Both types of process consist of a horizontal shaft with a series of discs attached to it and the discs are immersed in a tank containing sewage or settled sewage (Figs 5.1 and 5.2). The diameters of the discs can be up to 4 m and the shafts can be as long as 7 m, depending on the manufacturer. The units are covered to protect the biofilm which grows on the segmented plastic discs to a thickness of about 5 mm. Covering the units has the additional benefit of reducing heat losses and noise, of eliminating fly nuisance and, to some extent, of controlling odour problems. With RBCs, the depth of immersion of the discs is about 40% so that, as they rotate, the biofilm experiences a sequential sorption of BOD and ammoniacal-nitrogen from the liquid and oxygen from the atmosphere. The speed of rotation is around 1 r/min. The design loading rates tend to be specific to manufacturers, but the usual practice is that the BOD load per unit area and the quality of final effluent required are used

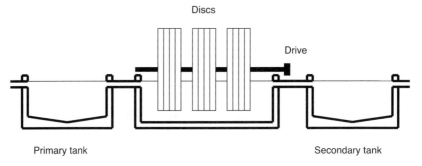

Fig. 5.1. Schematic diagram of a modular rotating biological contactor

to derive the disc area, which is based on both sides of the disc. The load should be derived from the population being served and the daily per capita mass of BOD. The RBCs produced by one company use a loading rate of $4\,\mathrm{g\,BOD/m^2}$ of disc area per day when an effluent quality of 25:45 is required and one of $2\,\mathrm{g\,BOD/m^2}$ per day when an ammoniacal-nitrogen concentration of 10 mg/litre or less is also required (Findlay, 2002).

Rotating biological contactors have been used for some time (Pike, 1978; Lumbers, 1983), but their operation has been plagued by mechanical failures in the rotor shafts, the media panels and the rods supporting the media (Brenner and Opaken, 1984; Mba *et al.*, 1999a). Recent work by Severn Trent Water and Cranfield University (Mba *et al.*, 1999b) has

Fig. 5.2. RBC unit (courtesy of Copa Ltd)

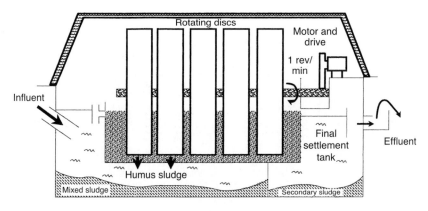

Fig. 5.3. Schematic diagram of an integral rotating biological contactor (Upton et al., 1995) (with permission from CIWEM)

resulted in a re-design of the overall system. This has dealt with the weak areas and resulted in a process which has a minimum life of 20 years. Rotating biological contactors can be operated as either modular or integral units. The modular approach requires the use of both primary and secondary settlement and tends to be used when the population is greater than 1000 (Upton *et al.*, 1995). The integral unit treats unsettled sewage and has the capability of providing both primary and secondary settlement within the unit (Fig. 5.3).

Figure 5.4 shows the performance of two RBC plants. Works A, which serves a population of 372, has both primary and secondary settlement tanks, whereas Works B merely has a settlement zone after the biodiscs and has a design load of 19 kg BOD per day. Both works are required to meet a consent of a BOD concentration of 30 mg/litre and a suspended solids' concentration of 40 mg/litre.

Submerged biological contactors differ from RBCs in that their discs have a greater immersion; typically 80–90% of the area is kept below the liquid surface. In addition, oxygen is supplied by two rows of diffusers. One supplies the needs of the biofilm, the process air, while the other rotates the shaft by being captured in peripheral cups (Figs 5.5 and 5.6). A typical split between the process air and the 'drive' air is 40:60 (Farrell *et al.*, 2001). Submerged biological contactor installations can be large. The unit at Fleetwood, UK, for example, has a flow to full treatment of 192 Ml/day. The use of step feeding has also been shown to be a useful way of maintaining a fully aerobic biofilm and avoiding the growth of species such as *Beggiatoa*, which can

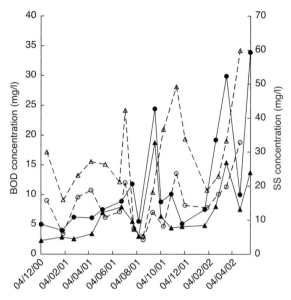

Fig. 5.4. Performance of two RBC plants: Works A — BOD (●) and SS (▲), and Works B — BOD (○) and SS (△)

produce sulphide. At the works at Morecambe, UK, 40% of the flow is introduced into the second stage (Farrell *et al.*, 2001). Thus, the plant has a total BOD (TBOD) load of 12 g/m^2 per day and an overall loading of 6·5 g TBOD/m^2 per day.

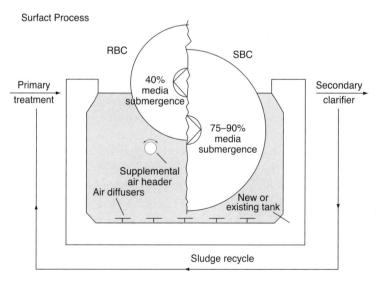

Fig. 5.5. Comparison of RBC and SBC layout

Bio-oxidation — other aspects

Fig. 5.6. Installation of SBC discs

Submerged filters

The origin of the submerged filter dates back to the beginning of the twentieth century when tanks containing submerged layers of slate were used, together with secondary settlement, to treat wastewater. The modern process, which consists of a submerged packed bed of granular media and which is aerated at the base of the bed (Fig. 5.7), was first introduced in the late 1970s.

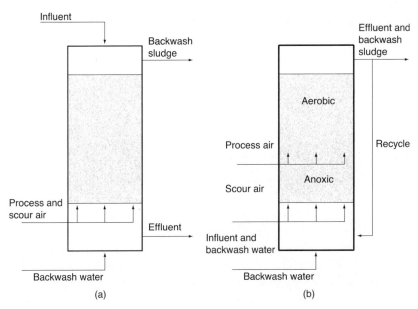

Fig. 5.7. Schematic diagram of BAF: (a) downflow; (b) upflow

117

Wastewater treatment and technology

Biological aerated filters

The main benefit of this process is that it occupies only a small area and is, thus, deemed to have a small 'footprint' (Smith et al., 1990). The wastewater being treated may flow upwards or downwards. This is a design decision and the choice depends on a number of factors. In the downflow mode, the maximum flow-rate will be dependent on the bed height and the pressure drop across the bed. This is not the case for upflow biological aerated filters (BAFs), where the maximum flow-rate is controlled purely by the pump capacity. If nitrification, as well as carbonaceous BOD removal, is required, it is claimed that a downflow reactor would be better (Vedry et al., 1994). A range of support media are used and may be structured, sunken or floating (Fig. 5.8). The materials used (Table 5.1) range from expanded shale (Biocarbone) to polystyrene (Biostyr). During operation, the granular media used in floating and sunken BAF configurations becomes clogged by biofilm growth and entrained solids, and must, therefore, be cleaned by regular backwashing (Park and Ganczarczyk, 1994). This may also involve an air scour (Fig. 5.7). The amount of backwash water is generally 2–15% of the total water treated (Carrand et al., 1990) and, typically, it removes about 30% of the biomass which is not expected to affect the performance of the reactor (Bacquet et al., 1991). Structured media, because of their open structure, require only occasional backwashing. The frequency of backwashing can be determined by one of three methods

- A regular timed basis, for example, every 24 hours. This is the simplest way and requires no additional equipment, but the time between backwashes may not be the optimum time;
- When a pre-set head loss occurs. This takes into account the variability of the sewage strength and, therefore, the biofilm growth;
- By monitoring effluent quality.

The factors to be considered in selecting the medium include its shape, size, surface texture and density. The density will affect the flow-rate, the backwashing rate and the capacity of the filter (Toettrup et al., 1994). As with the trickling filter (Chapter 3), the surface texture will affect the surface area for colonization and the attachment of the biofilm. A recent study (Nasr, 2001), which compared a sunken shale medium, had a relatively smooth surface, with a floating plastic medium, which was quite corrugated and showed intense colonization within the ridges of the latter medium (Fig. 5.9).

Bio-oxidation — other aspects

(a)

(b)

Fig. 5.8. (a) Shale and (b) plastic media used in biological aerated filters (Nasr, 2001)

Table 5.1. Examples of media used in BAFs

Process	Flow regime	Media	Media regime
Biobead	Upflow	Polyethylene	Floating
Biostyr	Upflow	Polystyrene	Floating
Biocarbone	Downflow	Expanded shale	Sunken
SAFe	Downflow	Expanded shale	Sunken
Biofor	Upflow	Biolite	Sunken
Biopur	Downflow	Polystyrene	Structured
Colox	Upflow	Sand	Sunken

119

Wastewater treatment and technology

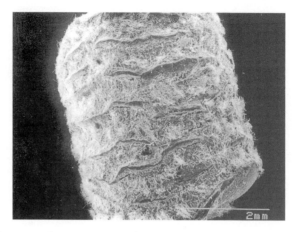

Fig. 5.9. *Scanning electron micrograph of colonized plastic from a BAF plant*

The loading rates which apply to BAFs will vary with the type of treatment required. Biological aerated filters can be used for the removal of carbonaceous BOD or they can be used specifically as nitrifying filters. As can be seen from Table 5.2, the loading rates, which have been applied to full-scale reactors, can be appreciable. These loads need to be compared with those used for trickling filters (0·06–0·12 kg BOD/m^3 per day) and for activated sludge plants (0·16–0·32 kg BOD/m^3 per day).

As with all types of wastewater treatment process, a BAF will give a performance which is largely in line with its original design and this will be based on the consent for the discharge. A recent review has reported

Table 5.2. *Loading rates applied to full-scale BAF plants*

Type of treatment	Loading rate	Source
Carbonaceous removal (kg COD/m^3 per day)	<3·25	Carrand et al., 1990
	0·83–2·65[a]	Smith and Hardy, 1992
	4·8–8	Pujol et al., 1994
	0·5–6·3	Pujol et al., 1994
	1·2[a]	Meaney, 1994
Nitrification (kg amm.N/m^3 per day)	0·38–0·76	Rogalla and Payraudeau, 1988
	<0·46	Carrand et al., 1990
	<0·8	Paffoni et al., 1990
	<0·25	Meaney, 1994
	0·19–0·35	Smith and Hardy, 1992
	<0·85	Vedry et al., 1994

[a] BOD loading

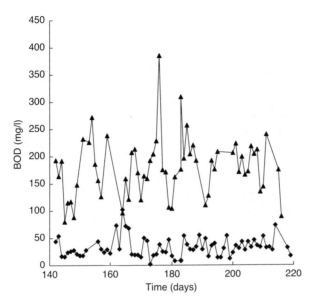

Fig. 5.10. Performance of a full-scale BAF plant showing the influent (▲) and effluent (◆) BOD

that the normally expected removal rates are 4·1 kg BOD/m^3 per day, 1·27 kg ammoniacal-nitrogen/m^3 per day and 5 kg nitrate-nitrogen/m^3 per day (Mendoza-Espinosa and Stephenson, 1999). Figure 5.10 shows a typical performance for a BAF plant required to oxidize carbonaceous pollutants. A similar performance was reported for a BAF plant using the SAFe media (Robinson et al., 1994). This treated settled sewage with a maximum flow of 14 000 m^3/d. The packing had a grain size of 2–6 mm and a specific surface of 100 m^2/m^3. Headloss was used to bring in the backwash stage which uses water and air scour. The backwash liquors contain a suspended solids' concentration of 1000–1800 mg/litre. The settled sewage, which was fed to the BAF plant, had a mean suspended solids' concentration of 80 mg/litre, a mean BOD of 75 mg/litre and a mean ammoniacal-nitrogen concentration of 13 mg/litre. The effluent produced by the plant had 95 percentile concentrations of BOD = 6·5 mg/litre, SS = 16 mg/litre and ammoniacal-nitrogen = 0·80 mg/litre.

The use of BAFs is not restricted to the treatment of domestic sewage — they are used to treat industrial wastewaters (Chapter 10). For example, a BAF has been used to treat a 50 m^3/d flow of brewery wastewater and achieved a COD removal of more than 90% at a loading rate of 6·25 kg COD/m^3 per day. The BAF technology has also

Table 5.3. Performance of full-scale B2A plant (Bigot et al., 1999)

	COD	NH_3-N	NO_3-N
Inlet concentration (mg/litre)	300–600	35–70	–
Inlet load (kg/m^3 per day)	2–14	0·4–1·1	–
Effluent concentration (mg/litre)	25–40	0–4	10·15

been used as a pre-treatment for the wastewater from a sulphite mill, achieving a mean BOD removal of 82% at a loading rate of 3·4 kg BOD/m^3 per day. The pre-treated effluent was then polished in an aerated lagoon (Kantardjieff and Jones, 1997).

If the process air is introduced not at the base of the filter but in the bed itself (Fig. 5.7(b)) and effluent is recycled, the anoxic zone at the base of the filter can effect de-nitrification. This has been done with the B2A process which uses a dual media packing (Bigot et al., 1999). With this system, a coarse medium (pumice, 4·6 mm) is used at the bottom of the filter and a fine medium (expanded clay, 3·3 mm) in the upper portion. The air is introduced at the interface of the two types of medium. A full-scale plant with this configuration has been in operation since 1992. Its performance is summarized in Table 5.3.

Submerged aerated filters

Submerged aerated filters (SAFs) are really plastic media filters (Chapter 3) which are operated in a way similar to a BAF. The plastic media which is used has a high voidage, more than 90%, and, therefore, does not require backwashing. However, it does not have the same straining effect as the media used in BAFs and the process should incorporate a sludge settlement stage.

Deep shaft process

This could be described as yet another variation on the activated sludge theme. It consists of a shaft in the ground. Depths range from 30–150 m and diameters from 0·4–3·4 m. The shaft is divided into upflow (riser) and downflow (downcomer) sections and the MLSS are circulated within the shaft by air-lift. The shaft is surmounted by a gas disengagement header tank. The sludge and the treated effluent are separated by settlement and the thickened sludge is returned to the header tank in the same way as in the activated sludge process. The mixed liquor leaving the header tank is supersaturated with nitrogen, carbon dioxide and

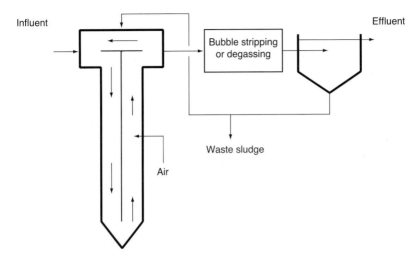

Fig. 5.11. *Schematic diagram of the deep shaft process*

oxygen as a result of the air having been dissolved under pressure in the lower regions of the shaft. Settlement can, therefore, be impeded by the presence of residual microbubbles. Because of this, when the liquors leave the header tank they pass to a de-gassing unit before being settled (Fig. 5.11). The de-gassing can be done by using a small coarse bubble aeration tank or vacuum de-gasser. Start-up is accomplished by injecting air into the riser limb. As the bubbles move to the surface, air-lift principles cause the liquors to start to circulate (Fig. 5.12(a)). When the fluid velocity in the downcomer exceeds the upflow velocity of the bubbles (typically about 300 mm/s) the process air is admitted into the downcomer and the start-up air can be either shut off or maintained at a flow rate to stabilize the flow regime (Fig. 5.12(b)). The movement of the liquid and air and the positioning of the process air inlet means that there are different fluid densities in the two limbs and it is this difference which maintains the circulation pattern. The turbulent flow (Reynolds Number greater than 100 000) which ensues gives very good mixing of the gas, the liquors and the biomass. The depth of the shaft means that the air is dissolved under pressure, the magnitude of which will depend on the depth of the shaft, so that oxygen transfer is high, of the order of 3 kg/m^3 per hour. This needs to be compared with the typical values for an activated sludge aeration tank of 0·1–0·3 kg/m^3 per hour. This, in turn, means that high MLSS concentrations can be used, 6–10 g/litre, and high F:M ratios (perhaps up to 1·8 kg BOD/kg MLSS per day) can be employed. Again, both these figures need to be compared with the values normally used for the activated sludge process (Chapter 4). The

Wastewater treatment and technology

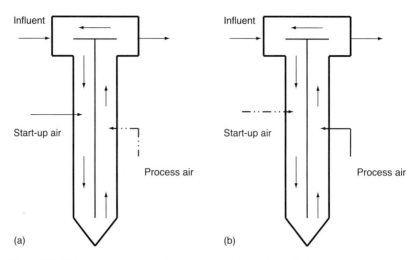

Fig. 5.12. Schematic diagram of the start-up of the deep shaft process

liquors take 2–5 minutes to circulate round the shaft and the hydraulic retention time is 20–40 circulations. This, together with the intense mixing within the shaft, means that it has a high resistance to toxic shock and changes in flow-rates or loading. It is also a process which is 'environmentally friendly'. Its land requirements are low, it is less obtrusive than conventional plant in that it is largely underground and it is claimed that odour and noise nuisances are lower. A further advantage is that the process can be used without primary treatment. The sludge which is produced by the deep shaft process is very similar to that in a mature activated sludge system. There is one exception — filamentous bacteria are seldom found. Indeed, it has been reported that even when a deep shaft plant was seeded with a filamentous sludge from a conventional activated sludge plant, the filaments disappeared within 12 hours (Hobbs et al., 1988).

It is a process which can be used for the treatment of both municipal sewage and industrial wastewaters. The deep shaft process is used at Tilbury, UK, to treat the flow from a design population equivalent of 325 000, and has a maximum flow of 72 000 m^3/d. The average flow is 30 000 m^3/d and the incoming BOD is 595 mg/litre (Irwin et al., 1989; Boon and Thomas, 1998). The shaft has an internal diameter of 5·7 m and a depth of 60 m. The header tank measures 32 × 8·25 m, with a water depth of 4 m. De-gassing is achieved by a coarse bubble aeration tank. This plant has been supplemented by a second deep shaft so that the size of the plant has effectively been doubled.

Little information is available about the design criteria for deep shaft systems. One way of dealing with this is to calculate the overall volume

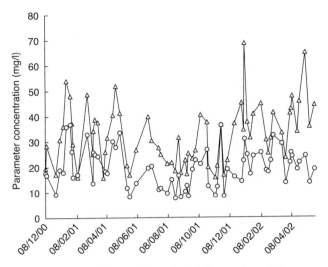

Fig. 5.13. *Final effluent quality (BOD ○; SS △) from a deep shaft treatment works*

(V in m^3) of a deep shaft system, including the header tank, from

$$V = \frac{L \times 24}{R \times M} \tag{5.1}$$

where L = design load based on the maximum BOD after balancing (kg/h), M = MLSS (g/litre), R = F:M ratio. The shaft depth is a balance between cost and oxygen transfer efficiency; a deeper shaft will cost more to construct but will give a higher transfer efficiency. The header tank does not exceed 25% of the overall volume and its plan area (m^2) is given by

$$A = \frac{A_d \times V_d}{U_b} \tag{5.2}$$

where A_d = cross-sectional area of downcomer (m^2), V_d = liquid velocity in downcomer (1–2 m/s), U_b = bubble rise rate (about 0·25 m/s). The minimum depth of the header tank would be 2 m with a freeboard of 0·5 m.

The location of the process air injection point will vary with the depth of the shaft. For depths of more than 100 m it would be at 35–45% of the hydraulic depth, for shafts with depths of about 50 m it would be 25–35%. At the Tilbury works, the downcomer injection point is at 47% of the hydraulic depth and the riser injection at 37%.

Figure 5.13 shows the effluent quality of a deep shaft treatment plant in terms of the BOD and suspended solids' concentrations. However, as

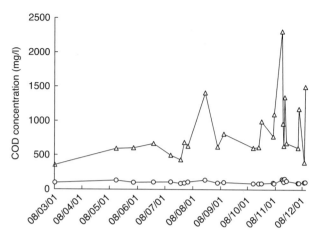

Fig. 5.14. *COD removal by a deep shaft treatment works (influent △; effluent ○)*

can be seen from Fig. 5.14, there is an appreciable amount of COD in the raw wastewater, much of it 'hard'. This means that the UWWT Directive and its requirements must also be considered.

Sequencing batch reactors

In effect, a sequencing batch reactor (SBR) system is an activated sludge process which is operated as a batch process rather than a continuous one. Originally, it was known as a fill and draw process and, in its simplest form, it operates in four phases (Fig. 5.15).

The single tank can, therefore, be thought of as acting as

- selector, i.e. the fill phase;
- the aeration tank, i.e. the aerated period;
- the final clarifier, i.e. the settlement and decant periods.

Also, it is possible to vary the 'aeration tank volume' by having a series of top water level (TWL) positions. The design criteria must reflect these factors. Thus

$$\text{OLR} = \frac{BQ}{(\text{MLSS at TWL}) \times (V \times \% \text{ aeration})} \tag{5.3}$$

$$\text{HRT} = \frac{V \times \% \text{ aeration}}{Q} \tag{5.4}$$

$$\text{sludge age} = \frac{(\text{MLSS at TWL}) \times (V \times \% \text{ aeration})}{\text{sludge wasted}} \tag{5.5}$$

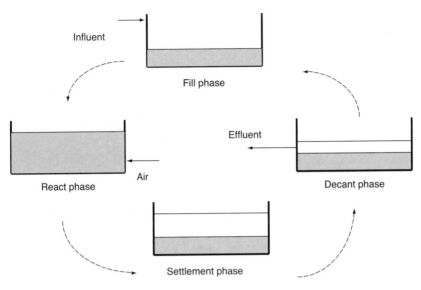

Fig. 5.15. Schematic diagram of the SBR treatment cycle

where HRT is the hydraulic retention time, % aeration is the proportion of time that the reactor is aerated, and the sludge age is in days. There are also a variety of ways in which the overall cycle of fill-aerate-settle-decant is achieved (Morgenroth and Wilderer, 1998). The sequence which is used will depend on the type of treatment required. Thus, for COD removal a simple fill, aerate, settle and decant sequence would be used. This also permits nitrification if the sludge age is long enough to allow a strong nitrifying population to develop. If de-nitrification is required an unaerated mixing stage is introduced between the fill and the aeration stages. Indeed, a series of fill, mix and aerate sequences might be used before the sludge was allowed to settle. The aeration of SBRs can be done with fine bubble diffusors, with surface aerators or by jets (Chapter 4). Surface rotors used at low speed would provide unaerated mixing, as would submerged jets when the oxygen/air supply was turned off. The other alternative for mixing would be to use a submerged propellor (Morgenroth and Wilderer, 1998).

The decant system is devised to prevent MLSS entering the effluent pipe-work during the fill and aeration periods, and prevent floating scum contaminating the final effluent.

This can be done with lift-clear decanters (Figs 5.16 and 5.17), or floating decanters with exclusion devices (Fig. 5.16).

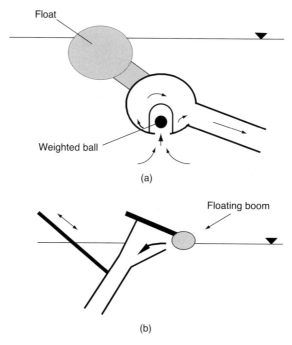

Fig. 5.16. Examples of decanters used in SBR plants. (a) Kvaerner floating decanter, and (b) CASS^{TM} lift-clear decanter

Although the concept of fill and draw reactors, which can be considered as the precursors of SBRs, is an old one it is only over the last decade that the use of SBRs has become readily accepted. The gradually increasing use of this technology can be seen from the data in Fig. 5.18. Some specifications for SBRs operated by Yorkshire Water for the treatment of municipal sewage are shown in Table 5.4 (Newbery and Young, 2000). They can also be used for the treatment of industrial wastewaters and Table 5.5 shows the performance of an SBR plant treating the liquors from a waste management company. These liquors had been pre-treated by primary settlement, oil separation and dissolved air flotation (Flapper et al., 2001).

Use of pure oxygen

Although air is used for most aeration systems, pure oxygen or oxygen enriched air can also be used. There are a variety of ways of contacting the oxygen with the mixed liquors and the settled sewage, but in this book examples of using open and closed tanks will be discussed.

(a)

(b)

Fig. 5.17. (a) SBR aeration tanks and (b) detail showing the aerators and the lift-clear decanter (Shandong Lukang Pharmaceutical Company, China — 4 Basin CASS^{TM} SBR system)

Wastewater treatment and technology

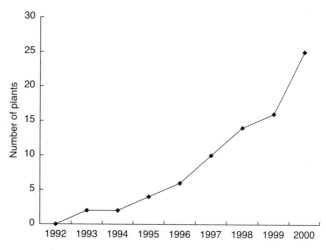

Fig. 5.18. SBR plants in Bavaria (Teichgräber et al., 2001)

Open tanks

Vitox™ and Oxy-Dep™ are aeration systems based on the use of pure oxygen and can be used either in a sidestream injection mode or as a skid-mounted unit located in the aeration tank. As a sidestream

Table 5.4. Characteristics of SBR plants operated by Yorkshire Water (Newbery and Young, 2000)

	Driffield	Brighouse	Beverley
Population equivalent	13 490	111 130	45 640
Upstream treatment	6 mm screens de-gritted	6 mm screens de-gritted and settled	6 mm screens de-gritted and settled roughing biofilter settlement
Consent			
BOD (mg/litre)	15	20	40
COD (mg/litre)	125	125	125
SS (mg/litre)	35	30	60
amm.N (mg/litre)	5	8	10
no. basins	4	4	2
shape	circular	rectangular	rectangular
vol. to TWL (m^3)	2088	9115	4906
depth to TWL (m)	6·0	6·0	5·0
decant depth (m)	1·7	1·35/1·55	1·7
cycle time (h)	4	4	4
% aeration	50	50	50
type aeration	FBDA	FBDA	FBDA

Table 5.5. *Influent and effluent characteristics of an SBR plant treating industrial wastewater (Flapper et al., 2001)*

	Influent (mg/litre)			Effluent (mg/litre)		
	Maximum	Minimum	Mean	Maximum	Minimum	Mean
BOD	30 000	4700	12 000	3000	40	500
SS	1600	120	700	1000	20	200
Phenol	48	1	25	10	–	5
Formaldehyde	53	2	4	10	–	1
TKN	210	40	120	40	5	30

injection process it is simple to install and operate. It requires

- a supply of oxygen;
- pipework;
- a pump;
- a venturi.

Settled sewage is pumped through a venturi where oxygen is injected under pressure (Fig. 5.19). The oxygenated sewage, which has a high

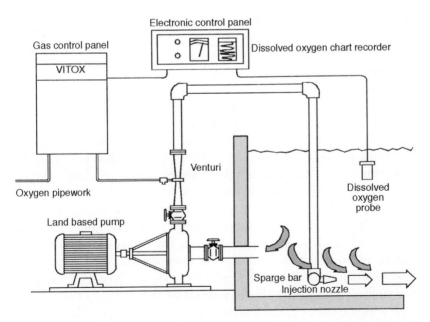

Fig. 5.19. *Schematic diagram of a Vitox sidestream unit (courtesy of BOC)*

Fig. 5.20. Skid mounted Oxy-Dep^{TM} unit (courtesy of Air Products Ltd)

concentration of dissolved oxygen (20–30 mg/litre), together with fine bubbles of undissolved oxygen, is then dispersed into the mixed liquor through a multi-nozzle sparge system. It would be usual to make this addition at the inlet end of a plug-flow aeration tank to achieve the best use of the oxygen. The skid-mounted unit (Fig. 5.20) works on a similar principle, but has a submersible pump as an integral part of the skid assembly.

These can be used either as the only aeration system or as a supplement (e.g. to cope with overloads for a specific period). One of the most rigorously analysed Vitox^{TM} installations is the Holdenhurst sewage treatment works (Robinson *et al.*, 1982; Toms and Booth, 1982). At this plant oxygen was used to uprate the treatment capacity. Oxygen can also be used in the treatment of industrial wastewaters. For example, the Conoco Refinery wastewater treatment plant consists of two parallel activated sludge units (4500 m^3). The aeration is provided by Vitox^{TM}. Each aeration lane contains two Vitox^{TM} aerators, each delivering 5 tonnes/day. The start-up flow was 4000 m^3/day and the BOD was reduced to less than 10 mg/litre. Drop-in Oxy-Dep^{TM} aerators give similar performances. For example, two drop-in Oxy-Dep^{TM} units, each with a capacity of 2·5 tonnes/day, were used to provide the aeration for the treatment of landfill leachate. The leachate had a strength

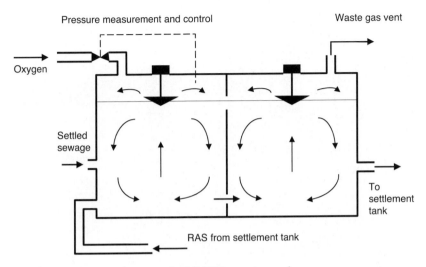

Fig. 5.21. Schematic diagram of a UNOX aeration tank

which varied between 5000 and 50 000 mg COD/litre, but the aerated lagoon consistently achieved a reduction of 95–98%.

Covered aeration tanks

With the UNOX process the aeration tank is covered and surface aerators, which drive through a water seal, operate in a pure oxygen (or oxygen enriched) environment. The pressure in this gas space is maintained at 20–50 mm water gauge (Figs 5.21 and 5.22). Oxygen transfer is greater than with air and, therefore, the process can accommodate a greater oxygen removal rate. This means that there can be a higher MLSS and, therefore, a higher sludge loading rate and a more compact plant. It also has the advantages of (Downing and Boon, 1983)

- giving improved settleability (Fig. 5.23);
- producing less sludge;
- reducing the risk of odours when dealing with a potentially malodorous wastewater.

The process can produce a good quality effluent in terms of BOD removal (Fig. 5.24) (Boon and Thomas, 1998). However, carbon dioxide can build-up in the liquors and depress the pH. This can affect nitrification. Although the UNOX process is widely used throughout the world (Downs, 1989; Kawata and Sakata, 1998), it

Wastewater treatment and technology

Fig. 5.22. General view of a UNOX aeration tank

has only achieved a very limited application in the UK, and only four plants have been built. This has been ascribed to a number of reasons, including the settlement tank design and the method of controlling the vented gases. An appraisal by Boon and Thomas (1998) examined the design of the final settlement tanks for the four UK plants and showed that all but one had been designed to operate with overflow velocities, which were greater than or equal to the critical value of 36 m^3/m^2 of plan area of the settlement tank per day. This meant that even with low SSVI values, there was insufficient floor area to achieve an adequate solid/liquid separation.

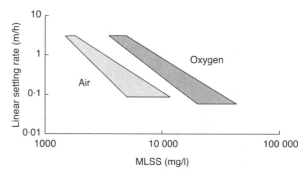

Fig. 5.23. Settlement characteristics for air and oxygen activated sludge

Bio-oxidation — other aspects

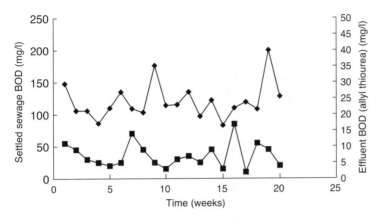

Fig. 5.24. Performance of a UNOX activated sludge system showing the influent (♦) and effluent (▲) BOD concentrations

Oxygen supply

The oxygen required by these processes can be supplied in one of three ways. It can be generated on-site cryogenically, it can be stored on-site as liquid oxygen (Fig. 5.25) or it can be generated on-site by swing

Fig. 5.25. Oxygen storage facility

Wastewater treatment and technology

Fig. 5.26. Pressure swing adsorption facility

adsorption using either pressure or vacuum. With swing adsorption a molecular sieve is used which is able to adsorb nitrogen and carbon dioxide from the air at a specified pressure or vacuum, giving a supply of oxygen-rich air (Fig. 5.26). The sieve can be re-generated by returning the pressure to atmospheric pressure. Each method of supply has a cost associated with it and, therefore, the choice of which method to use is

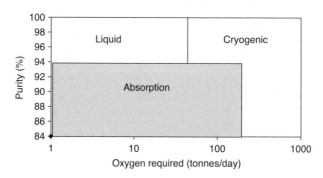

Fig. 5.27. Options for the supply of oxygen

site-specific and will depend on the amount of oxygen required, and the length of time for which the oxygen would be used. Thus, for example, if sidestream injection were being used to cope with a seasonal overload, liquid oxygen would probably be the preferred choice. Figure 5.27 provides a guide to the choice of supply, but no more than a guide.

Oxidation ditches

Although the oxidation ditch is not strictly a modern variant, it is frequently discussed as a separate process (Barnes *et al.*, 1983). It was originally developed in Holland as a cheap, rural, batch process. Essentially, it was and is an activated sludge process which uses an aeration tank, having the form of a 'continuous racetrack' and a horizontal shaft (paddle) aerator (the TNO rotor) with a diameter of 0·75 m (Figs 5.28 and 5.29). These aerate the mixed liquors and propel them round the ditch with a velocity of 0·25–0·30 m/s. This is sufficient to keep the activated sludge in suspension. They can achieve an oxygen transfer of the order of 2–3 kg oxygen/metre length of rotor per hour. The number of rotors used will depend on the amount of oxygen needed to treat the sewage to the required standard. The original concept was for it to operate as a fill and draw unit, with the ditch acting as both an aeration tank and a settlement tank. The philosophy became popular and the system was developed into a continuous process, and a separate settlement tank was provided.

Fig. 5.28. TNO rotor

Wastewater treatment and technology

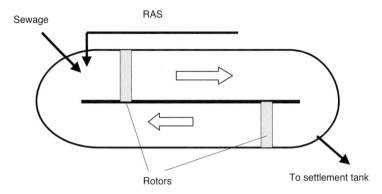

Fig. 5.29. Schematic diagram of an oxidation ditch

The ditches operate at low organic loading rates, typically in the range 0·05–0·08 kg BOD/kg MLSS per day. This is in the extended aeration range and it is not unusual to operate with hydraulic retention times of 24 hours or even longer. The TNO rotors cannot mix the aeration tank liquors adequately at depths greater than 1·8 m. Therefore, the land requirements for large population equivalents is excessive. Oxidation ditches are definitely a 'high footprint' process.

Their success has led to alternative aeration systems being developed which can operate in deeper (3–3·5 m) aeration tanks with increased oxygen transfer. The two main developments have been the Carrousel system (Fig. 5.30) which uses vertical shaft aerators (see Fig. 4.9), and Mammoth aeration (Figs 5.31 and 5.32) which uses a larger horizontal shaft rotor (1·0 m diameter) with a modified blade configuration. This can transfer up to 8 kg oxygen/metre of rotor per hour with an efficiency of 2·0–2·5 kg O_2/kWh. These deep oxidation ditches are not always operated as extended aeration systems. A survey of Mammoth aeration plants showed hydraulic retention times as low as three hours (Forster, 1980).

The time for the liquors to complete a single circuit of a ditch will be measured in minutes. Therefore, the mixing characteristics of oxidation ditches can be considered in one of two ways. In the short term they have a flow which is approaching plug-flow, but, taken over the hydraulic retention time, they are completely mixed. This has to be taken into consideration when any mathematical modelling is being attempted. It also means that oxidation ditches are well able to withstand shock loads because of the high dilution which occurs.

Because it is an extended aeration system, the endogenous oxygen requirement (2×10^{-3}MV in equation (4.13)) is appreciable so that

Bio-oxidation — other aspects

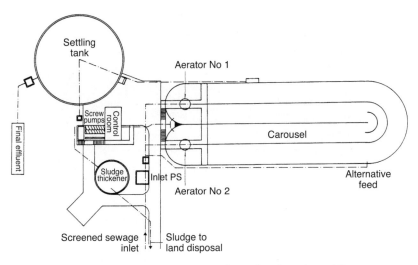

Fig. 5.30. *Layout of the Ash Vale Carrousel (Rachwal et al., 1983)*

the process efficiency (kWh/kg BOD removed) is higher than that of a conventional plant. The result is that oxidation ditches are said to be expensive to operate. In fact, some of the oxygen/energy used during the endogenous phase is being used to stabilize the sludge and, therefore, there will be lower running costs for sludge processing.

Other types of aeration can be used in oxidation ditches (Forster, 1983). Jet aeration, which uses an ejector system where mixed liquor is entrained with air, is one of these. A manifold of aeration jets is arranged across the bottom of the ditch, and a horizontal velocity is

Fig. 5.31. *Mammoth rotor (courtesy of Whitehead and Poole Ltd)*

Wastewater treatment and technology

Fig. 5.32. Mammoth oxidation ditch in operation

imparted to the liquors by the thrust of the jet and a vertical mixing occurs by the rise of the plume.

Oxidation ditches can also be operated in a multi-channel mode. This is the Orbal system (Drews *et al.*, 1972; Applegate *et al.*, 1980; Burrows *et al.*, 2000). As is shown in Figs 5.33 and 5.34, the settled sewage enters the outer channel and then flows sequentially to the inner channels. Aeration, mixing and flow within the channels are achieved by perforated discs (1·2 cm thick × 1·3 m diameter).

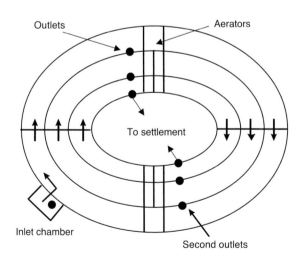

Fig. 5.33. Schematic diagram of an Orbal activated sludge system

Bio-oxidation — other aspects

Fig. 5.34. Orbal aeration tank showing the multiple lanes and the (covered) aerators

Another approach is to use ditches operated in pairs or in triplicate with no final settlement tank and no return sludge pumping. In effect, each ditch operates, in turn, as a sequencing batch reactor, usually with a de-nitrification phase (Bundgard and Holm Kristensen, 1982). This variant has very good nitrification/de-nitrification characteristics. The plant discussed by Bundgard and Holm Kristensen achieved a BOD removal of 98·8% and a total nitrogen removal of 86·6%.

Oxidation ditches can achieve BOD removal, ammoniacal-nitrogen oxidation and de-nitrification. In other words, they can produce a very good quality effluent (Fig. 5.35 and Table 5.6). De-nitrification depends on there being an anoxic regime within the aeration tank. This can occur naturally if the aerobic respiration of the sludge reduces the dissolved oxygen to zero. This will tend to happen because of the

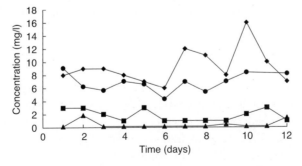

Fig. 5.35. Variations in TSS (◆), BOD (■), ammoniacal-nitrogen (▲) and nitrate-nitrogen (●) in the effluent from an Orbal oxidation ditch

141

Table 5.6. *Final effluent quality from oxidation ditches*

Type	Mean effluent concentrations (mg/litre)			
	BOD	SS	Amm.N	TON
Simple ditch	5	10	0·7	12
Simple ditch	7	14	3	17
Mammoth	9·3	16·7	1·2	20·6
Mammoth	12·8	14·3	1·8	25·8
Orbal	1·8	9·3	0·4	6·8
Carrousel	2·6	7·1	1·3	6·4
Carrousel	2·8	3·9	1·5	3·7

TON = total oxidized nitrogen

ditch configuration, the spatial arrangement of the aerators and the high circulation rate (Johnstone *et al.*, 1983; Lin and Sah, 2002).

Reed beds

Reed beds or constructed wetlands were introduced into the UK in 1985 and have considerable potential for treating the flows from small or isolated communities. The examination of this potential led to guidelines for their design and operation being introduced in 1990 (Cooper, 1990) and modified in 1996 (Cooper *et al.*, 1996). As well as being able to treat sewage, they are very environmentally friendly and provide an important haven for wildlife (Fig. 5.36). Although other species can be used, the most commonly used species is *Phragmites australis*, planted, as seedlings, at a density of about 4 plants/m^2 of the bed. The reeds have three roles

- they host bacterial colonies around the roots (aerobic, anoxic, anaerobic) of around 10^{10}–10^{11} m/litre of the volume of the soil around the roots;
- they assist the transport of oxygen in the root zone;
- the rhizospheres which develop act as hydraulic transport pathways.

The soil acts to supply the bacterial colonies and buffer against sudden effluent increases. In practice, the reeds can be established in either a soil or a gravel bed, but it is more common to use gravel (6–10 mm diameter) because it provides hydraulic conductivity without relying on the root systems (Cooper, 2001). In either case the matrix is retained by an impermeable membrane such as clay or a synthetic liner. Reed beds can be constructed as horizontal-flow or vertical-flow systems (see Fig. 7.11).

Bio-oxidation — other aspects

Fig. 5.36. *An operational reed bed (courtesy of Severn Trent Water)*

Horizontal-flow beds have a base with a slope of 2–8%. The inlet zone is comprised of a bed of small stones which support the distribution pipe. The bed has an average depth of 0·6 m and is 10–15 m long. The outlet zone is constructed in a similar way to the inlet zone, and the treated effluent is collected in perforated pipes which discharge to a level control device which maintains the water level at a pre-determined point below the surface of the gravel. Initially, this is set at about 50 mm below the surface of the gravel, but is lowered to ensure that there is sub-surface flow as the hydraulic conductivity of the bed decreases with time (Griffin and Pamplin, 1998).

Reed beds are good at removing suspended solids and bacteria as they have an appreciable ability to filter. They are also able to denitrify very effectively. However, their oxygen transfer capability is limited and, therefore, this controls the degree of BOD removal and means that nitrification in horizontal-flow beds is poor. However, this is not critical when the beds are being used for tertiary treatment (Chapter 7). The design of horizontal-flow beds for secondary treatment is based on 5–10 m²/pe. In the UK the area (A in m³) is calculated from

$$A = \frac{Q(\ln C_o - \ln C_t)}{k} \qquad (5.6)$$

where Q = average flow (m³/d), C_o = average BOD of the feed (mg/litre), C_t = required average BOD of the effluent (mg/litre), k = rate constant taken as 0·06–0·1 (Cooper 2001). Vertical-flow beds are flat beds of graded gravel (5–10 mm) topped with sand, and are dosed in pulses so that each pulse floods the surface and then percolates down through the bed. This means that there is good aeration as each pulse also carries oxygen down into the bed. The BOD removal and nitrification are, therefore, good. Suspended solids are collected on the surface and may cause clogging. The design of vertical-flow beds for secondary treatment is based on 1 m²/pe for BOD removal and 2 m²/pe for BOD removal and nitrification (Cooper, 2001). An important aspect in the design of vertical flow beds is to consider the risk of the top layer of sand becoming clogged by solids. One way of minimizing this is to use at least four beds whose use can be rotated, thus enabling beds to rest (Cooper, 1999). Restricting the organic loading rate to 25 g COD/m² per day will also help to minimize the impact of clogging (Platzer and Mauch, 1997).

These two types of reed bed can be used in combination and may be used with either the vertical-flow or the horizontal-flow stage placed first. Results from a plant using a horizontal-flow bed as the initial stage show that the system could achieve a good removal of BOD and suspended solids. However, it has been recommended that, for secondary treatment, it is better to use vertical flow beds as the initial stage because they can oxidize BOD and ammoniacal-nitrogen as well as remove bacteria (Cooper, 1999). O'Hogain and Gray (2002) describe the use of this latter configuration to treat the outflow from a septic tank of 12 m³/d. The design used two vertical-flow stages followed by a horizontal-flow stage. The primary vertical-flow bed was operated as four discrete units in parallel which were operated in rotation to allow beds to rest. The cycle used was two days in operation and six days resting. The results show that the removal of BOD, averaged over a period of two years, was 91%, the suspended solids' removal was 99% and that of ammoniacal-nitrogen was 67%.

Severn Trent Water have a policy of using reed beds for treating flows from small communities, which are defined as less than 2000 pe, and where the consent standards are more stringent than 25 mg/litre BOD and 45 mg/litre suspended solids; the reed bed would be preceded by a rotating biological contactor (Griffin and Upton, 1999).

Membrane reactors

Membrane bioreactors use membranes instead of a final clarifier to effect the solids/liquid separation stage. Essentially, two types of reactor design can be identified. One uses the membrane in a side-stream mode outside the aeration tank and returns concentrated sludge in the same way as would a conventional activated sludge plant. The other has the membrane unit submerged in the aeration tank (Fig. 5.37). Each option has advantages and disadvantages (Table 5.7), all of which need to be considered for any particular application. Although ceramic membranes are available, polymeric materials, such as acrylonitrite, polyethylene and polysulphone, are

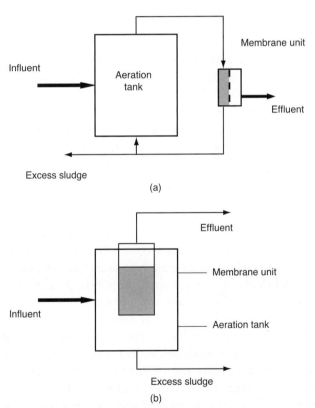

Fig. 5.37. Schematic diagram of membrane bioreactor systems with: (a) an external membrane system and (b) a submerged membrane

Table 5.7. *Advantages and disadvantages of MBR configuration (Till and Mallia, 2001)*

Submerged MBR	Sidestream MBR
High aeration costs	Low aeration costs
Low liquid pumping costs	High liquid pumping costs
Lower flux and higher footprint	Higher flux and lower footprint
Less frequent cleaning	More frequent cleaning
Lower operating costs	Higher operating costs
Higher capital costs	Lower capital costs

more likely to be used. The membranes generally have pore sizes of 0·1 to 0·5 μm and may be configured as plate and frame, hollow fibre or tubular units (Gander et al., 2000). In terms of operation, consideration has to be given to the flux, that is, the flow rate per unit area of membrane, the trans membrane pressure, which is the driving force, and how these may vary with time. It is also necessary to consider cleaning of the membranes. Although the flux will be affected by factors such as the quality of the feed and the configuration, material and area of the membrane, the critical factor is membrane fouling. This means that cleaning is also critical. With submerged systems, the scouring action of the aeration is sufficient to ensure that any more thorough cleaning, for example with hypochlorite, is required infrequently, perhaps only twice a year. Typical flux values are 0·5–1·0 m^3/m^2 per day and the trans membrane pressures could be as low as 0·1 bar (10^4 Pa).

These plants can be operated at very high MLSS concentrations, typically 16–20 g/litre, with F:M ratios of 0·02–0·04 and relatively long sludge ages. This means that they can be very compact. It also means that with the long sludge ages there is an appreciable amount of endogenous respiration which leads to a low sludge production, less than 0·3 kg/kg BOD (Mayhew and Stephenson, 1997; Gander et al., 2000). An additional benefit is that there is an appreciable removal of bacteria and viruses (Ueda and Horan, 2000). The performance of membrane bioreactors has been examined by a number of workers (Gander et al., 2000; Lozier, 2000). One such set of data which compares the performance of a membrane bioreactor system with that of a conventional activated sludge plant is shown in Table 5.8. The microbiological descriptions of the sludge which develops in submerged membrane reactors suggest that, in general, the sludge is different from that in conventional activated sludge plants (Rosenberger et al., 2000;

Table 5.8. A comparison of the performances of a membrane bioreactor system and a conventional activated sludge plant operating in parallel (Lozier, 2000)

	Membrane reactor	Activated sludge plant
MLSS (mg/litre)	15 119	3241
Hydraulic retention time (h)	3·61	29·8
Sludge age (d)	3·61	12·4
Influent		
BOD (mg/litre)	189	189
Amm.N (mg/litre)	27·6	27·6
SS (mg/litre)	203	203
Effluent		
BOD (mg/litre)	0·55	1·32
Amm.N (mg/litre)	3·08	0·24
SS (mg/litre)	0·6	not reported

Wagner and Rosenwinkel, 2000). There tends to be a predominance of freely suspended bacteria and small aggregates or micro-colonies. Another feature is the low number of protozoans. Perhaps the only real similarity with conventional activated sludge was the occurrence of filamentous species. Overall, it was concluded that changes in the sludge structure had no effect on the plant performance (Rosenberger et al., 2000). This is a very positive aspect of membrane reactors as it means that bulking and foaming would cease to be a problem.

Although a number of companies supply membrane bio-reactor (MBR) systems, the two main suppliers are Kubota (Japan) and Zenon (USA). The main characteristics of the two processes are summarized in Table 5.9.

Table 5.9. Characteristics of the main MBR systems

	Supplier	
	Kubota	Zenon
Membrane material	Polyolefin	
Membrane configuration	Flat sheet	Hollow fibre
Nominal pore size (μm)	0·4	0·4
Flux (litre/m^2 per hour)	20	40–70
Trans-membrane pressure	0·1 bar	10–50 kPa
Typical MLSS (g/litre)	15–20	15–20

Membranes can also be used to separate the solid and liquid phases as part of an anaerobic digestion process (Fakru'l-Razi and Noor, 1999; Wen et al., 1999). Using a membrane in conjunction with a contact digester would avoid having to use a de-gassing system prior to the settlement tank (see Fig. 9.3).

Odours and odour control

General overview

Odours can occur at any type of wastewater treatment plant and the control of odours can cause appreciable costs, both capital and operating (Vincent and Hobson, 1998). Odour nuisances occur when

- compounds which have an odour are formed or are present in the incoming sewage;
- there is a transfer of odours from the liquid phase to the air;
- the air moves to a populated area.

Many compounds can be classified as being odorous and can be present in sewage as microbial breakdown products or as inputs from industry. Probably the most common source of odour associated with sewage treatment is hydrogen sulphide, H_2S. This is a toxic, odorous and corrosive gas which is formed from sulphate, SO_4, under anaerobic conditions by the sulphate reducing bacteria. Some sulphate in sewage will have originated from the raw water and some of it from human excreta (about $0.7-1.1\,g\,SO_4$/capita per day).

Most compounds which smell have published odour thresholds (Table 5.10). Their recognizable concentrations are greater than the odour thresholds, with the nuisance concentrations being about 5–10 times greater. The pH can affect liquid/air partitioning. Thus, low pH values will favour the emission of H_2S and volatile fatty acids, while high pH values will favour emissions of ammonia and amines.

Table 5.10. Examples of odour threshold values

Compound	Smell	Odour threshold (ppb)
Hydrogen sulphide	Rotten eggs	0·50
Dimethyl sulphide	Rotten vegetables	0·12–0·40
Skatole	Faecal	0·002–0·060
Butyric acid	Rancid	0·09–20

Table 5.11. *Suggested exposure limits (Vincent and Hobson, 1998)*

Compound	Long-term (8 h) exposure limit (ppm)	Short-term (10 min) exposure limit (ppm)
Hydrogen sulphide	10	15
Ethyl mercaptan	0·5	2
Methyl amine	10	–

In 2003, the legislation relating to odours is contained in The Environmental Protection Act, 1990, which defines a statutory nuisance as

> 'any dust, steam, smell or other effluvia arising on industrial, trade or business premises and being prejudicial to health or a nuisance'

The legislation is enforced by Environmental Health Officers who can issue abatement orders, and failure to comply could result in fines (up to £20 000).

In 2003, there are no mandatory European or UK standards, but it is possible that there will be in-house standards for specific works. Works personnel, as well as the general public, need to be considered (Control of Substances Hazardous to Health Regulations) and, therefore, occupational exposure limits are likely to be specified (Table 5.11).

It should be noted that

- The maximum concentration to which the general public should be exposed is of the order of 1/40 of the occupational exposure limit (OEL), but, for hydrogen sulphide, this would be 0·25 ppm and, thus, would not be acceptable in odour terms;
- H_2S is a toxic gas, and 1000–2000 ppm would give immediate collapse and paralysis of respiration. Even at concentrations of 350–500 ppm there is a risk of death;
- The typical background concentration of H_2S in the UK is 0·5–1·5 parts per billion (ppb), where billion means 1000 million.

Odour measurement

The overall measurement of odours requires the human nose. Odorous air is diluted until 50% of a test panel (eight people) can detect no odour. The number of volumes of clean air used per unit volume of sample is the odour concentration (odour units are designated ou), with units of ou/m^3. The limit of sensitivity is of the order of 20–50 ou/m^3. Background odours are often in the region of 20–100 ou/m^3 and nuisance

Table 5.12. *Typical odour concentrations at a sewage works (Vincent and Hobson, 1998)*

Location	Typical concentration ranges	
	Odour (ou/m^3)	H$_2$S (ppm)
Screening plant	100–5000	1–10
De-sludging chamber	1000–10 000	10–100
Primary sludge tank	1000–500 000	10–500
Sludge de-watering plant	1000–50 000	0·5–50

limits can be 5–10 ou/m^3, so it is not an over-accurate technique for tracking down odours.

Hydrogen sulphide is likely to be a major part of a 'sewage works smell' and it can be used as an 'indicator' smell. It can be measured instrumentally to 1 ppb (threshold odour concentration = 0·5 ppb) and, therefore, be used to 'map' a works, and can be measured in both air and liquid phases and the results used to model its release. Both these approaches can be used to identify 'hot spots' at a works. Typical odour concentrations at various locations at a sewage works are given in Table 5.12. Alternative techniques for other components are available, but they are more complex and, therefore, more expensive.

Dealing with odours

Odours can be handled both at the design stage and during plant operation. What is necessary is the avoidance of 'septicity' or regions, where the redox potential is such that H$_2$S formation will be promoted. This can occur in:

(a) Sewers

Essentially this means in rising mains and the concentration of sulphide can be calculated from (Boon and Lister, 1975)

$$C_S = K_C t COD[(1 + 0·004D)/D]1·07^{(T-20)} \tag{5.7}$$

where C_S = sulphide concentration (mg S/litre), K_C = constant (0·00152), t = anaerobic retention time (min), D = rising main diameter (cm), COD = sewage COD (mg/litre), T = temperature (°C). The formation of H$_2$S in sewers will occur in pumped or rising mains, where there is no air/water interface and, therefore, no natural re-aeration. Under

Table 5.13. The effect of temperature on the oxygen demand of sewage and sewer slime

Temperature (°C)	Oxygen demand of sewage (mg/litre per hour)	Oxygen demand of slime layer (mg/litre per hour)
11 i.e. winter	4·8	500
21 i.e. summer	9·5	1000

these circumstances the biofilm on the sewer wall and, to some extent, the biomass in the sewage will quickly utilize any dissolved oxygen in the sewage (Table 5.13), and generate the anaerobic conditions necessary for the formation of H_2S. These effects can be modelled (Spooner, 2001), thus enabling engineers to develop odour control measures. The effects can be reduced by minimizing the retention times where anaerobic conditions can prevail, i.e. minimizing the lengths of pumped sections. An alternative approach is to add chemicals which can either maintain aerobic conditions in the pumped sections or oxidize any H_2S which is formed. The injection of oxygen into rising mains has been used successfully to control H_2S odours (Boon *et al.*, 1977). Other chemicals which have been used include hydrogen peroxide, chlorine and nitrate (Sims, 1980; Jefferson *et al.*, 2002). As well as creating odour problems, hydrogen sulphide can be oxidized by bacteria to sulphuric acid and cause corrosion in sewers.

(b) Inlet works
If H_2S has been generated within the sewerage system, odours can be emitted at the inlet works as the sewage comes into the atmosphere. However, dissolved concentrations of less than 0·25 mg/litre are unlikely to give problems and release can be minimized by reducing turbulence in the septic sewage.

Remedial measures include dosing at the inlet works with iron salts or hydrogen peroxide (Sims, 1980; Jefferson *et al.*, 2002), air stripping at the inlet works, although this will mean that the waste air will require treatment, and using covered tanks and sumps which are then vented to an odour control unit.

(c) Main treatment works
Four points need to be considered. These are

Wastewater treatment and technology

- The proximity of housing, shops etc. likely to be affected by any odours should be examined to determine the sensitivity of the site. This could affect the odour standard for the site;
- At the design stage, processes which are less likely to produce odours or which can be covered easily should be chosen. For example, the high-rate plastic filter (Chapter 3) with an upflow of air has the potential for giving odours, whereas using a co-current flow of liquid and air allows any odours to be contained;
- Potentially smelly processes should be arranged so that a centralized odour treatment facility can be provided. Another alternative is to consider using odorous air as a process air for another process, e.g. combustion air for an incinerator (Chapter 8);
- The handling/processing of raw sludge will create odours. Therefore, these processes should be covered and the air extracted for treatment.

Most of these procedures can be incorporated at the design stage of a new works or a works extension (Drust and Deacon, 1995). This may be done by applying no more than common sense in deciding which areas are likely to pose problems (Fig. 5.38).

In the event of there being complaints associated with odours it is essential to identify the sources. This should include identifying any background smell from other industry in the vicinity of the sewage works. Depending on what sources are identified, it may be necessary

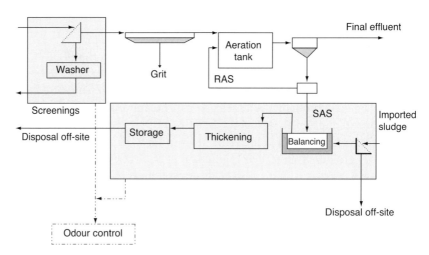

Fig. 5.38. Schematic diagram of a sewage treatment plant showing the areas which could be enclosed to minimize the effects of odours

Fig. 5.39. *The use of biofilters for odour control*

to control septicity in sewers or tanks, instigate treatment, either with chemicals or by stripping, or seal/cover open areas such as high-rate filters, channels and tanks.

There are several methods which can be used to treat odiferous air:

(a) *Biofilters and bioscrubbers.* Biofilters (Fig. 5.39) are best for H_2S concentrations of less than 15 ppm and are beds of moist fibrous material such as peat or coir. Bioscrubbers will treat higher concentrations and use a support medium of plastic, mussel shells or calcified seaweed. The bacteria in the biofilm which develops, oxidize H_2S to H_2SO_4. This can cause the pH to drop and the calcium in the mussel shells or calcified seaweed will neutralize this.

(b) *Wet chemical scrubbers.* These are essentially standard gas cleaning scrubbers. They use oxidants (peroxide, ozone), acid and an alkali for H_2S (Drust and Deacon, 1995).

(c) *Catalytic oxidation.* Iron will react with hydrogen sulphide to form iron sulphide and ultimately elemental sulphur:

iron oxide + H_2S → iron sulphide → iron oxide

+ elemental sulphur

These reactions are exothermal and release about 7·1 kJ/g sulphur produced. If this approach is used in the form of a filter, it is likely that micro-organisms will grow on the iron support and be able to oxidize organic sulphur compounds. Boon and Boon

(1999) have described the initial performance of three full-scale 'rusty iron' filters using sections of mild steel pipe (15 mm long × 25 mm diameter) as the packing. The results indicate that this is a successful way of removing sulphide with the performance dependent on the inlet load and the air flow conditions. The sulphur which will accumulate in this type of filter could be removed by steam cleaning and would require a disposal route. This could be to the chemical industry or to a suitable land fill site.

(d) *Dry scrubbing/adsorption.* This method can use activated carbon or media impregnated with an oxidizing agent such as potassium permanganate. It is used to treat vented air and will have a finite life. There must, therefore, be regular replacement of the media.

As a general rule, the choice will depend on the size of the problem. That is to say, the flow-rate of air requiring treatment. Thus, as a rule of thumb

- Intermittent flows — adsorption with activated carbon;
- $<1000\, m^3/h$ — adsorption with activated carbon;
- $1000-10\,000\, m^3/h$ — biofilters;
- $>10\,000\, m^3/h$ — chemical scrubbing.

(e) *Chemical oxidation.* During the recent up-grading of one sewage treatment works (STW), the odour control measures included covering the inlet works (i.e. screening and screenings handling and grit removal) and the sludge sumps from the primary tanks. The vented gas is then passed through a tower system packed with 'pumice' onto which chlorine dioxide, which is a strong oxidant, has been sorbed. The life of the pumice/chlorine dioxide will depend on the amount of H_2S being emitted, but a five-year life is being claimed.

(f) *De-odorants.* De-odorizer sprays have been used for some time to mask sewage works odours and have been successful in reducing complaints, but they should not be considered as a long-term solution. There have also been occasions when the de-odorizer smell has been considered to be worse than the original one. The use of de-odorants should be undertaken in collaboration with works personnel and the general public likely to be affected.

References

Applegate, C. S., Wilder, B. and Deshaw, J. R. (1980) Total nitrogen removal in a multi-channel oxidation system, *J. Wat. Pollut. Control Fed.*, **52**, 568–77.

Bacquet, G., Joret, J. C., Rogalla, F. and Bourbigot, M. M. (1991) Biofilm start-up and control in aerated filters, *Environ. Technol.*, **12**, 747–56.

Barnes, D., Forster, C. F. and Johnstone, D. W. M. (1983) *Oxidation Ditches in Wastewater Treatment*, Pitman Books, London.

Bigot, M., Le Tallec, X. and Badard, M. (1999) A new generation of biological aerated filters, *J. CIWEM*, **13**, 363–8.

Boon, A. G. and Boon, K. (1999) Catalytic-iron filters for effective and low-cost treatment of odorous air, *J. CIWEM*, **13**, 189–95.

Boon, A. G. and Lister, A. R. (1975) Formation of sulphide in a rising-main sewer and its prevention by injection of oxygen, *Prog. Wat. Technol.*, **7(2)**, 289–300.

Boon, A. G. and Thomas, V. K. (1998) Intensification of the activated sludge process, *J. CIWEM*, **12**, 357–68.

Boon, A. G., Skellett, C. F., Newcombe, S., Jones, J. G. and Forster, C. F. (1977) The use of oxygen to treat sewage in a rising main, *Wat. Pollut. Control*, **76**, 98–112.

Brenner, R. C. and Opaken, E. J. (1984) Design information on rotating biological contactor design, *Technical Report Data*, Municipal Environmental Research Laboratory, Cincinnati, Ohio.

Bundgard, E. and Holm Kristensen, G. (1982) The operation of oxidation ditches including the Bio-Denitro system in Denmark, *Proc. Int. Conf. Oxidation Ditch Technology*, Amsterdam, Oct., CEP Consultants, Edinburgh, pp 117–128.

Burrows, L. J., West, J. R., Forster, C. F. and Martin, A. (2000) Mixing studies in an Orbal activated sludge system, *Water SA*, **27**, 79–83.

Carrand, G., Capon, B., Rasconi, A. and Brenner, R. (1990) Elimination of carbonaceous and nitrogenous pollutants by a two-stage fixed growth process, *Wat. Sci. Technol.*, **22(1/2)**, 261–72.

Cooper, P. F. (1990) *European Design and Operations Guidelines for Reed-bed Treatment Systems, WRc Report UI 17*, Swindon, UK.

Cooper, P. (1999) Design and performance of vertical-flow and hybrid reed bed treatment systems, *Wat. Sci. Technol.*, **40(3)**, 1–9.

Cooper, P. (2001) Constructed wetlands and reed-beds: Mature technology for the treatment of wastewaters from small populations, *J. CIWEM*, **15**, 79–85.

Cooper, P. F., Job, G. D., Green, M. B. and Shutes, R. B. E. (1996) *Reed-beds and Constructed Wetlands for Wastewater Treatment*, WRc, Swindon.

Downing, A. L. and Boon, A. G. (1983) High intensity systems of activated sludge, in *Comprehensive Biotechnology, Part 3*, Robinson, C. W. and Howell, J. A. (eds), Pergamon Press, Oxford, pp 881–98.

Downs, T. (1989) Tough Canadian environmental standards being met by Howe Sound, *Pulp and Paper*, **63(9)**, 132–3.

Drews, J. L. C., Malan, W. M., Meiring, P. G. J. and Moffat, B. (1972) The Orbal extended aeration activated sludge plant, *J. Wat. Pollut. Control Fed.*, **42**, 221–31.

Drust, J. and Deacon, R. (1995) A case history of odour control using chemical scrubbers, *J. CIWEM*, **9**, 199–206.

Fakru'l-Razi, A. and Noor, M. J. M. M. (1999) Treatment of palm oil effluent (POME) with the membrane anaerobic system (MAS), *Wat. Sci. Technol.*, **39(10–11)**, 159–72.

Farrell, P. D., Nair, A., Palmer, S. and Davies, W. J. (2001) Control and stabilisation of odorous sludge from Fleetwood sewage treatment works, *J. CIWEM*, **15**, 46–50.

Findlay, G. E. (2002) Personal communication.

Flapper, T. G., Ashbolt, N. J., Lee, A. T. and O'Neill, M. (2001) From the lab to full-scale SBR operation: treating high strength and variable industrial wastewater, *Wat. Sci. Technol.*, **43(3)**, 347–54.

Forster, C. F. (1980) A comparison of the performances achieved by the Carrousel and the Mammoth rotor versions of the oxidation ditch, *Environ. Technol. Letters*, **1**, 366–75.

Forster, C. F. (1983) The historical development of oxidation ditches, in *Oxidation Ditches in Wastewater Treatment*, Barnes, D., Forster, C. F. and Johnstone, D. W. M. (eds), Pitman Books, London, pp 1–15.

Gander, M., Jefferson, B. and Judd, S. (2000) Aerobic MBRs for domestic wastewater treatment: a review with cost considerations, *Sep. Purif. Technol.*, **18**, 119–30.

Griffin, P. and Pamplin, C. (1998) The advantages of a constructed reed bed based strategy for small sewage treatment works, *Wat. Sci. Technol.*, **38(3)**, 143–50.

Griffin, P. and Upton, J. (1999) Constructed wetlands: A strategy for sustainable wastewater treatment at small treatment works, *J. CIWEM*, **13**, 441–6.

Hobbs, M. G., Hobbs, J. R., Sasser, L. W. and Daly, P. G. (1988) Selecting and design of a deep shaft activated sludge facility for Homer, Alaska, *Paper presented at WPCF 61st Annual Conference, Oct.*

Irwin, R. A., Brignal, W. J. and Biss, M. A. (1989) Experiences with the Deep-Shaft process at Tilbury, *J. IWEM*, **3**, 280–7.

Jefferson, B., Hurst, A., Stuetz, R. and Parsons, S. A. (2002) A comparison of chemical methods for the control of odours in wastewater, *Trans. I. Chem. E.*, **80 Part B**, 93–9.

Johnstone, D. W. M., Rachwal, A. J. and Hanbury, M. J. (1983) General aspects of the oxidation ditch process, in *Oxidation Ditches in Wastewater Treatment*, Barnes, D., Forster, C. F. and Johnstone, D. W. M. (eds), Pitman Books, London, pp 41–74.

Kantardjieff, A. and Jones, J. P. (1997) Practical experience with aerobic biofilters in TMP (thermomechanical pulping), sulfite and fine paper mills in Canada, *Wat. Sci. Technol.*, **35(2–3)**, 227–34.

Kawata, K. and Sakata, Y. (1998) Wastewater treatment in pulp and paper with UNOX System high purity oxygen activated sludge system, *Jap. Tappi J.*, **52(6)**, 78–84.

Lin, L. Y. and Sah, J. G. (2002) Nitrogen removal in leachate using Carrousel activated sludge treatment process, *J. Env. Sci. Health Part A — Tox./Haz. Subs. Env. Engng.*, **37**, 1607–20.

Lozier, J. (2000) Two approaches to indirect potable reuse using membrane technology, *Wat. Sci. Technol.*, **41(10–11)**, 149–56.

Lumbers, J. P. (1983) Rotating biological contactors: Current problems and potential developments in design and control, *The Public Health Engineer*, **11**, 41–5.

Mayhew, M. and Stephenson, T. (1997) Low biomass yield activated sludge: A review, *Env. Technol.*, **18**, 883–6.

Meaney, B. (1994) Operation of submerged filters by Anglian Water Services Ltd., *J. IWEM*, **8**, 327–34.

Mba, D., Bannister, R. H. and Findlay, G. E. (1999a) Condition monitoring of low-speed rotating machinery using stress waves: Part 1, *Proc. Instn. Mech. Engrs.*, **213**, 153–70.

Mba, D., Bannister, R. H. and Findlay, G. E. (1999b) Mechanical redesign of the rotating biological contactor, *Water Res.*, **33**, 3679–88.

Mendoza-Espinosa, L. and Stephenson, T. (1999) Review of biological aerated filters (BAFs) for wastewater treatment, *Env. Eng. Sci.*, **16**, 201–16.

Morgenroth, E. and Wilderer, P. A. (1998) Sequencing batch reactor technology: Concepts, design and experiences, *J. CIWEM*, **12**, 314–21.

Nasr, S. Z. (2001) *Characterization of Biofilms Associated with Sunken and Floating Biological Aerated Filters (BAFs)*, PhD Thesis, University of Birmingham.

Newbery, C. and Young, G. (2000) Sequencing batch reactors in Yorkshire Water, *Activated Sludge — The Lessons of the Last Decade*, CIWEM Conference, Nottingham, March.

O'Hogain, S. and Gray, N. F. (2002) Colecott hybrid reed-bed system: Design, construction and operation, *J. CIWEM*, **16**, 90–5.

Paffoni, C., Gousailles, M., Rogalla, F. and Gilles, P. (1990) Aerated filters for nitrification and effluent polishing, *Wat. Sci. Technol.*, **22(7/8)**, 181–9.

Park, J. W. and Ganczarczyk, J. J. (1994) Gravity separation of biomass washed out from an aerated submerged filter, *Environ. Technol.*, **22**, 181–9.

Pike, E. B. (1978) The design of percolating filters and rotary biological contactors, including details of international practice, *Technical Report, TR 93*, Water Research Centre, Medmenham.

Platzer, C. and Mauch, K. (1997) Soil clogging in vertical-flow reed beds — mechanisms, parameters, consequences and solutions, *Wat. Sci. Technol.*, **35(5)**, 175–81.

Pujol, R., Hamon, M., Kandel, X. and Lemmel, H. (1994) Biofilters: Flexible, reliable biological reactors, *Wat. Sci. Technol.*, **29(10/11)**, 33–8.

Rachwal, A. J., Johnstone, D. W. M., Hanbury, M. J. and Carmichael, W. F. (1983) An intensive evaluation of the Carrousel system, in *Oxidation Ditches in Wastewater Treatment*, Barnes, D., Johnstone, D. W. M. and Forster, C. F. (eds), Pitman Books, London, pp 132–72.

Robinson, A. B., Brignal, W. J. and Smith, A. J. (1994) Construction and operation of a submerged aerated filter sewage-treatment works, *J. IWEM*, **8**, 215–27.

Robinson, M., Varley, R. A. and Kimber, A. R. (1982) The use of oxygen to uprate the treatment capacity of a conventional surface-aeration plant at Holdenhurst (Bournemouth) sewage treatment works, *Wat. Pollut. Control*, **81**, 391–8.

Rogalla, F. and Payraudeau, M. (1988) Tertiary nitrification with fixed biomass reactors, *Water Supply*, **6**, 347–54.

Rosenberger, S., Witzig, R., Manz, W., Szewzyk, U. and Kraume, M. (2000) Operation of different membrane reactors: Experimental results and physiological state of the micro-organisms, *Wat. Sci. Technol.*, **41(10–11)**, 269–77.

Sims, A. F. E. (1980) Odour control with hydrogen peroxide, *Prog. Wat. Technol.*, **12(5)**, 609–20.

Smith, A. J. and Hardy, P. J. (1992) High-rate sewage treatment using biological aerated filters, *J. IWEM*, **6**, 179–93.

Smith, A. J., Quinn, J. J. and Hardy, P. J. (1990) The development of an aerated filter package plant, *Proc. 1st Int. Conf. Advances in Water Treat. Environ. Mngt.*, June, Lyon, France.

Spooner, S. B. (2001) Integrated monitoring and modelling of odours in sewerage systems, *J. CIWEM*, **15**, 217–22.

Teichgräber, B., Schreff, D., Ekkerlein, C. and Wilderer, P. A. (2001) SBR technology in Germany — an overview, *Wat. Sci. Technol.*, **43(3)**, 323–30.

Till, S. and Mallia, H. (2001) Membrane bioreactors: Wastewater treatment applications to achieve high quality effluent, *Paper presented at the 64th Annual Water Industry Engineers and Operators Conference*, Bendigo, September.

Toettrup, H., Rogalla, F., Vidal, A. and Harremoes, P. (1994) The treatment trilogy of floating filters: From pilot to prototype to plant, *Wat. Sci. Technol.*, **19(10/11)**, 23–32.

Toms, R. G. and Booth, M. (1982) The use of oxygen in sewage treatment works, *Wat. Pollut. Control*, **81**, 151–65.

Ueda, T. and Horan, N. (2000) Fate of indigenous bacteriophage in a membrane reactor, *Water Res.*, **34**, 2151–9.

Upton, J. E., Green, M. B. and Finlay, G. E. (1995) Sewage treatment for small communities: the Severn Trent approach, *J. CIWEM*, **9**, 64–71.

Vedry, B., Paffoni, C., Gousailles, M. and Bernard, C. (1994) First month's operation of two biofilter prototypes in the wastewater plant of Acheres, *Wat. Sci. Technol.*, **29(10/11)**, 39–46.

Vincent, A. and Hobson, J. (1998) *Odour Control*, CIWEM Monographs of Best Practice No. 2, CIWEM.

Wagner, J. and Rosenwinkel, K-H. (2000) Sludge production in membrane bioreactors under different conditions, *Wat. Sci. Technol.*, **41(10–11)**, 251–8.

Wen, C., Huang, X. and Qian, Y. (1999) Domestic wastewater treatment using and anaerobic bioreactor coupled with membrane filtration, *Process Biochem.*, **35**, 335–40.

6

Nutrient removal

Introduction

The accumulation of nitrogen and phosphorus compounds, particularly nitrates and phosphates, in natural waters will result in the waters becoming eutrophic. This will stimulate the growth of the primary producers, algae and plants. The UWWT Directive defines this as the 'accelerated growth of algae and higher forms of plant life to produce an undesirable disturbance to the balance of organisms present in the water and to the quality of the water concerned'. The precise concentrations which can be used to define eutrophication vary with the source of the data. For example, the UK Environment Agency has proposed that for standing waters an annual geometric mean of more than 85 µg P/litre would define eutrophic conditions, while the OECD has made the suggestion that a eutrophic water would contain an annual average concentration of 35–100 µg/litre of total phosphorus.

The UWWT Directive, therefore, specifies that there should be limits on the nutrient concentrations discharged to 'sensitive waters'. Alternatively, it requires a specified percentage reduction. These standards, essentially, relate to the larger treatment works and vary with the population equivalent (Table 6.1).

In the UK, three criteria are used to define sensitive waters

- that they are, or have the potential for being, eutrophic;
- that they contain, or could contain, if no action were taken, more than 50 mg nitrate/litre;
- that they are waters which are in need of protection to meet the requirements of other Directives.

Table 6.1. *Nutrient standards required by the UWWT Directive*

Works size (pe)	Total N		Total P	
	Concentration (mg/litre)	Minimum removal (%)	Concentration (mg/litre)	Minimum removal (%)
10 000–100 000	15	70–80	2	80
>100 000	10	70–80	1	80

At the end of 2001, 92 areas in the UK had been designated as being sensitive (www.defra.gov.uk/environment/water/quality/uwwtd).

Nitrogen removal

Nitrogen removal will involve the processes of nitrification and denitrification.

Nitrification

The process of nitrification is two-stage and, in the main, is carried out by the autotrophic bacteria *Nitrosomonas* (ammonia to nitrite) and *Nitrobacter* (nitrite to nitrate). However, modern techniques of genetic characterization are starting to show that other species may be involved:

$$NH_3 \rightarrow NO_2 \rightarrow NO_3$$
$$NO_3 + C \rightarrow N_2$$

Nitrification will occur in both the activated sludge process and trickling filter systems if they are designed properly (see Chapters 3 and 4). In trickling filters, there is what could be called eco-stratification with the upper regions being colonized by the heterotrophic carbon-oxidizers and nitrification occurring in the lower parts of the filter. Thus, if the applied organic loading rate becomes too high, the nitrifying species will be 'squeezed out' of the system due to the increased competition with the carbon-oxidizing species. A similar effect can be caused by increases in the hydraulic loading rate which will extend the growth of the heterotrophic bacteria throughout the bed. Hydraulic loading rates above $2 \cdot 5 \text{ m}^3/\text{m}^3$ per day are said virtually to eliminate nitrification.

For activated sludge, proper design, essentially, means that the process must be operated with a sludge age, which is long enough to maintain a strong population of nitrifying bacteria in the sludge. The nitrifying bacteria grow more slowly than the carbon-oxidizers. Typically, they have a maximum specific growth rate (μ_{max}) of 0·3–0·6 day^{-1} at 10 °C, compared with one of about 0·7 h^{-1} for those species which oxidize BOD. This means that their growth is more susceptible to temperature. It is not unknown for treatment plants to lose their ability to nitrify during the winter months or at least to have this ability severely impaired. To some extent, this can be counteracted by increasing the MLSS concentration during the winter months. Nitrification is also affected by pH. Taking both these points into consideration, it is possible to calculate the minimum sludge age (θ) necessary to sustain nitrification using equations of the type (Downing, 1968; Marais, 1973)

$$\frac{1}{\theta} = [0·18 - 0·15(7·2 - \text{pH})]\exp[0·12(T - 15)] \quad \text{see equation (4.2)}$$

$$\theta = 3·05(1·127)^{(20-T)} \tag{6.1}$$

Alternatively, performance curves of the type shown in Fig. 4.7 can be used. As a rule of thumb, at 10 °C, a sludge age of more than or equal to 10 days and a DO of more than or equal to 2 mg/litre ought to ensure good nitrification. If these conditions do not produce the nitrification required the settled sewage should be checked for alkalinity and the presence of toxin/inhibitory compounds. The process of nitrification uses alkalinity. The amount of alkalinity required will depend on the concentrations of ammoniacal-nitrogen present in the wastewater, but it can be expected that 1 g of ammoniacal-nitrogen will consume 6–7·2 g of alkalinity (Barnes and Bliss, 1983). It can be seen, therefore, that a lack of alkalinity may restrict nitrification in areas where the water is particularly soft.

Purely domestic sewage is unlikely to contain any inhibitory compounds, but trade effluent discharges may need to be screened since inhibitory compounds can be contained by seemingly inocuous discharges. For example, the effluent from a photographic processing plant was found to contain thiourea at a sufficiently high concentration to have a serious impact on the nitrification at the wastewater treatment works treating it (Hayes, 1999).

When filters are used specifically for nitrifying, the ammoniacal-nitrogen loads, as well as the BOD loads, must be considered. Most UK installations operate at loadings of 0·06–0·09 kg ammoniacal-nitrogen/m^3 of packing media per day, and at the lower end of this

Table 6.2. *Loading rates for nitrifying filters*

Maximum loading rate		Settled sewage	
(kg BOD/m^3 per day)	(kg amm.N/m^3 per day)	BOD (mg/litre)	amm.N (mg/litre)
0·15	0·02	<110	<20
0·12	0·01	<150	<25

range should achieve 90% nitrification. However, if better than 5 mg/litre, on a 95 percentile basis, is required the loading rates specified in Table 6.2 need to be applied.

In addition, the following operational conditions should be applied

- temperature >7 °C;
- pH = 7·0–8·5;
- ratio of peak:mean flows not greater than 2·0;
- deeper filters: 2·5–3·0 m.

Nitrifying filters (Fig. 6.1) would use a finer media than that used in conventional trickling filters (Chapter 3). Typically, a medium with a diameter of 20–30 mm and specific surface of 200–250 m^2/m^3 would be suitable. Modular plastic media can also be used for nitrifying

Fig. 6.1. *Nitrifying filters being used after an activated sludge plant (courtesy of Severn Trent Water)*

Wastewater treatment and technology

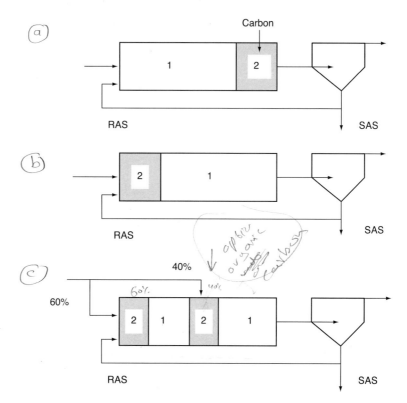

Fig. 6.2. De-nitrification options (1 aerobic zone; 2 anoxic zone)

filters. Whichever type of media is used, it is essential that there is good ventilation because of the high oxygen requirement for the process, i.e. 4·3 g O_2/g ammoniacal-nitrogen oxidized.

De-nitrification

De-nitrification requires an anoxic zone and a carbon source, with the bacteria in the sludge taking the oxygen they require for respiration and utilization of the carbonaceous material from the nitrate. This can be done (Fig. 6.2) (Cooper et al., 1994)

- Immediately before the final settling tank. At this stage the BOD would be very low and, therefore, supplementary carbon needs to be added;
- Immediately prior to or at the beginning of the aeration stage. The carbon required for the process is the BOD in the settled sewage;

Table 6.3. Activated sludge performance, February 2000 to January 2001 (Robinson, 2001)

Site	Estimated total-N (mg/litre)	Estimated removal of total-N (%)
C	16·3	55
F	23·8	36
M	14·1	53
S-1	13·2	63
S-2	19·8	43

- As a two-stage process, with a proportion of the settled sewage being fed to an anoxic region within the aeration lane.

The process of de-nitrification is less reliable. Estimates of the performance of large works in one water company's region have shown that most of them would not meet the requirement of 10 mg total nitrogen/litre (Table 6.3).

Most activated sludge plants in the UK which practise de-nitrification do so in the form of pre-de-nitrification (Fig. 6.2). That is to say, the return activated sludge is contacted with the settled sewage in an anoxic reactor or zone prior to entering the main aeration stage, and uses the incoming BOD as the carbon source for the de-nitrification process. The hydraulic retention time in this anoxic region will, therefore, be the time available for de-nitrification, and the rate of this process must be such that there is an adequate removal of nitrate. This pre-supposes that there is sufficient carbon in the settled sewage, and there is a large enough population of bacteria capable of carrying out the de-nitrification.

It has been acknowledged that there are difficulties of operation and design of anoxic zones and that achieving the required standards will cause problems (Upton et al., 1993).

Some bacterial strains are more efficient at achieving de-nitrification than others (Magnusson et al., 1998) and the specific de-nitrification rates can vary from works to works (Naidoo et al., 1998). Batch tests have shown that there are two, three or, at some works, four different rates (Naidoo et al., 1998; Robinson, 2001). These different rates have been shown to be associated with the utilization of the different forms of COD (Naidoo et al., 1998). The first rate is relatively fast and, with acetate as the carbon source, is in the range 3–7·3 mg N/g volatile suspended solids per hour (Naidoo et al., 1998). Recent work

Wastewater treatment and technology

Table 6.4. De-nitrification rates

Site	Date	First de-nitrification rate (mg/g SS per hour)	Second de-nitrification rate (mg/g SS per hour)
C	23-7-01	2·69	0·04
	9-8-01	3·22	1·73
D	26-7-01	3·38	2·03
	2-8-01	4·67	2·07
F	25-6-01	4·25	1·51
	3-7-01	3·24	1·35
M	24-7-01	7·7	2·62
	6-8-01	7·43	1·73
S-1	27-7-01	4·53	2·56
	7-8-01	4·32	3·31
S-2	31-7-01	7·86	2·73
	8-8-01	1·88	0·17
W	25-7-01	6·86	2·55
	1-8-01	4·68	2·46

(Robinson, 2001) has also shown this and has demonstrated that these rates vary from day to day and from works to works (Table 6.4). The work also showed that the duration of the first rate was variable (Fig. 6.3).

If the fast first rates could be achieved consistently with settled sewage, full de-nitrification could be achieved with a hydraulic retention time in the anoxic zone of one hour, assuming a nitrate concentration of around 25 mg N/litre. The data in Fig. 6.3 show that this is not the case. However, the carbon source used will control the

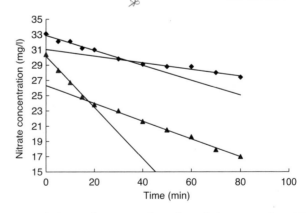

Fig. 6.3. Potential de-nitrification trials with settled sewage from Works S on 31-7-01 (▲) and 8-8-01 (●)

rate of de-nitrification. In one study it was found that while acetate gave good, rapid de-nitrification, the use of propionate resulted in no de-nitrification occurring (Takai et al., 1997). Another study showed differences between simple molecules, such as acetate and methanol, and more complex ones, such as crude syrup and hydrolyzed starch (Lee and Welander, 1996). The nature of settled sewage is complex, and probably the only parameter available to a works operator for judging what impact the composition of the feed to the anoxic zone might have on the de-nitrification process is readily biodegradable chemical oxygen demand (RBCOD). This is not a parameter which is normally measured. There are values in the literature for the various COD fractions. These are usually associated with mathematical modelling and do vary widely. Reported values for RBCOD range from 9% of the total COD (Kappeler and Gujer, 1992) to 20%, on average (Orhon and Ubay Çokgör, 1997), to 61·6% (Burrows, 2000). This range has been confirmed by a recent report which quotes values ranging from 10–35% of the total COD (Rössle and Pretorious, 2001). This wide variation means that there really is no way for a works operator to know or estimate how the incoming settled sewage might affect the anoxic process. It is also likely to vary from day to day and from season to season. At present (2003), there is no knowledge about this. In addition, settled sewage will contain components, organic or inorganic, which may inhibit the process of de-nitrification. For example, zinc has been shown to have a marked impact on the specific de-nitrification rate, with a concentration of 50 mg/litre causing a decrease from a value of 19·6 mg NO_3/g MLSS per hour to one of 5·3 mg NO_3/g MLSS per hour (Panswad and Polprucksa, 1998). Any attempt to optimize de-nitrification must, therefore, be based on a sound knowledge of the characteristics of the settled sewage, something which will be site-specific.

The anoxic zones at most sewage treatment works are uncovered and it is possible for oxygen to diffuse into the liquors. It has been shown that oxygen penetration has a severe impact on the process when de-nitrification rates are low (Jobbagy et al., 2000). Another study has shown that dissolved oxygen concentrations as low as 0·09 mg/litre can cause a decrease of 35% in the de-nitrification rate (Oh and Silverstein, 1999). Improvements to the de-nitrification rates by preventing the ingress of oxygen is a route which the water companies could follow, at a cost, but, without a better understanding of the factors which affect the de-nitrification rates, it would not necessarily guarantee the high-quality effluent required by the UWWT Directive.

The use of a fixed film de-nitrifying unit is an alternative to using a suspended growth anoxic zone, and deep bed filters (Chapter 7) have successfully been used for this purpose. The designs are similar to those of a rapid gravity filter although a coarser sand tends to be used. For example, with the Tetra filter a sand with a diameter of 2–3 mm is used compared with 1–2 mm for conventional filters (Upton *et al.*, 1993). Alternatively, an un-aerated version of a BAF (Chapter 5) could be used. De-nitrifying filters can be installed before the main aeration tank or after the final settlement tank (Upton *et al.*, 1993; Cooper *et al.*, 1994). In the latter case, it would require a carbon supplement. This could be ethanol, acetate or methanol. For methanol, the reaction is

$$5CH_3OH + 6NO_3^- = 5CO_2 + 3N_2 + 7H_2O + 6OH^-$$

This means that 1·9 mg of methanol are required to reduce 1 mg of nitrate. However, in practice, greater quantities are needed and work with a pilot-plant in the UK has reported a methanol requirement of 4·1–4·4 g/g total oxidized nitrogen removed (Upton *et al.*, 1993). Settled sewage can also be used as the carbon source and it has been reported that the requirement is 2·8 g BOD/g nitrate-nitrogen (Cooper and Wheeldon, 1980). The process of de-nitrification will result in biological growth and, therefore, the need for backwashing. The frequency of backwashing will depend on a range of factors such as filter medium, temperature and loading rate. The removal rates which can be obtained appear to be fairly consistent and are in the range 0·70–0·77 kg nitrate nitrogen/m^3 per day.

Phosphorus removal

Phosphorus (ortho-phosphate, organo-phosphates or polyphosphate) in sewage comes mainly from human excreta and from synthetic detergents (Fig. 6.4).

Phosphate can be removed from wastewaters chemically or biologically.

Chemical phosphorus removal

Chemical treatment involves adding calcium, iron or aluminium salts to form insoluble phosphates which are then removed by settlement. However, the removal chemistry is more complex than this and involves the formation of hydroxides as well as phosphates. Initially,

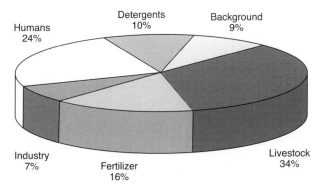

Fig. 6.4. Sources of phosphate in the aqueous environment

hexaqua ions are formed:

$$Fe^{3+} + 6(H_2O) = (Fe(H_2O)_6)^{3+}$$

The positive charges of the trivalent ions cause the bonds within the water molecules to polarize, leading to the liberation of protons and the formation of an insoluble hydrated metal hydroxide

$$(Fe(H_2O)_6)^{3+}{}_{(soluble)} = Fe(H_2O)_3(HO)_{3\,(precipitate)} + 3H^+$$

The reaction is pH dependent, with progressively more protons being released under alkaline conditions. It has been reported that iron (III) hydroxide has an Fe:OH ratio of 2·8:1 at pH 6·0 (He et al., 1995). This iron hydroxide has a highly disordered ramified chain structure and, viewed under a scanning electron microscope, each link was seen to be a quasi-spherical unit with a diameter of 1–4 nm (He et al., 1995). This precipitate is gelatinous and enmeshes any particles found in the sewage, including the precipitated iron phosphate

$$Fe^{3+} + PO_4^{3-} = FePO_4$$

The most common form of aluminium used to remove phosphate from wastewater is alum, $Al_2(SO_4)_3 \cdot 14H_2O$. It is thought that aluminium reacts with orthophosphate first, creating the insoluble aluminium phosphate (Tenney and Stumm, 1965), and only when most of the phosphate is precipitated (i.e. less than 10 mg PO_4/litre) will the trivalent reaction similar to that for iron occur (Hsu, 1975). When aluminium is added as the sulphate, the wastewater may need some additional alkalinity because of the acidity from the sulphate. Aluminium salts are the most efficient chemicals for precipitating phosphorus and produce lower volumes of sludge in comparison with iron salts.

Wastewater treatment and technology

Table 6.5. *Phosphate removal from trickling filter effluents in the Lake District (Stocks et al., 1994)*

	Influent (mg TP/litre)	Effluent (mg TP/litre)	Removal (%)
Hawkeshead	5·4	0·61	89
Ambleside	6·2	1·01	84

With both aluminium or iron salts, coagulating aids, such as anionic polyelectrolytes, may be necessary to promote the removal of dispersed metal phosphate floc. Typical doses are 0·1–0·25 mg/litre (Brett et al., 1997).

With trickling filters (or other film-based processes) chemical treatment is the usual option. With ferric ions, the iron to phosphate ratio is between 6 and 7 and the point of addition is usually directly into the final (humus) tank. When consideration was being given to the removal of phosphorus from the trickling filters discharging into Lake Windermere, it was determined that aluminium would have resulted in a residual having the potential for fish toxicity and that lime would have given a high pH effluent and a high production of sludge. Therefore, the use of ferric sulphate with redox control was chosen (Stocks et al., 1994). The results of this treatment are shown in Table 6.5.

With activated sludge, chemicals can be added at any one of three points (Fig. 6.5)

- before the primary tank (pre-precipitation);
- into the aeration tank (simultaneous precipitation);
- after the final tank (post-precipitation).

Each has advantages and disadvantages.

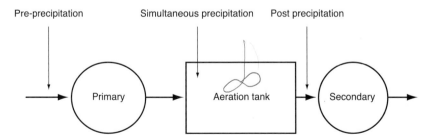

Fig. 6.5. *Dosing options for chemical removal of phosphate in the activated sludge process*

Pre-precipitation
(a) Phosphorus is removed in the primary sludge.
(b) The iron:phosphorus ratio (using ferric-iron) would be between 2:1 and 4:1.
(c) Fine suspended solids and BOD will also be precipitated and/or coagulated. This will result in a greater removal of these components in the primary tank than would normally occur. This will happen whether it is needed or not. Typically, the BOD removal would increase from about 30% (Chapter 2) to 50–70%.
(d) The coagulant demand associated with this removal of BOD and suspended solids will result in higher chemical doses than is required for phosphate removal and, therefore, has higher operational costs.
(e) The addition of chemicals at this point would only remove soluble ortho-phosphate. This is between 70 and 80% of the total phosphorus. Organo-phosphates and poly-phosphates (20–30% total phosphorus) would not have become hydrolysed and would not be removed.

Simultaneous precipitation
(a) Dosing into the aeration tank means that it is possible to use cheaper ferrous salts which would then be oxidized to the ferric state in the aeration tank. These are the more effective form. The dose would depend on the concentration of phosphate in the settled sewage with an iron:phosphorus ratio of 2:1 being typical.
(b) The oxidation of the ferrous salts will use oxygen, in theory $0.15\,g/g\,Fe^{2+}$. This will need to be considered in design calculations, but de Haas et al. (2001) have shown that the oxygen demand due to iron will only be a small part of the overall oxygen demand.
(c) Good mixing would be expected in the aeration tank and, therefore, there would be good contacting between the coagulant and the ortho-phosphate.
(d) Bioflocculation would enhance the precipitation and, hence, lower the doses required.
(e) The presence of the coagulant would improve the settlement characteristics of the activated sludge.
(f) This approach is the easiest to retro-fit.
(g) Much of the organo-phosphate and poly-phosphate would have been hydrolysed to ortho-phosphate by biological activity in the aeration tank and would, therefore, be removable.

Post precipitation

(a) Normal bio-oxidation processes will remove some ortho-phosphate through uptake by the biomass. Typically, this would be 10–15% for trickling filters and 20–40% for activated sludge.
(b) Most of the residual phosphorus (90–95%) is present as ortho-phosphate and, therefore, is capable of being precipitated.
(c) Most of the organic matter which could interfere has been removed.
(d) When ferric salts are used, the resultant floc is very light and there is a danger that solids can be carried over into the final effluent. This means that there is a need to de-sludge the final tanks frequently to prevent rising sludge and poor effluent quality.

Whichever option is selected, it must be recognized that there will be an increase in the production of waste sludge. It has been suggested that, with iron dosing, the average increase in volume will be as great as 40% (Upton, 1997). A survey of wastewater treatment plants in France showed that for a molar ratio of $Fe:P_{removed}$ of 1·5, the excess mineral sludge production was 3·95 kg total solids/kg $P_{influent}$, and that the total cost at that time associated with this route for phosphorus removal was 1·76 euros/kg of phosphorus in the influent (Paul et al., 2001).

Although iron salts are the most commonly used chemical, there are occasions (perhaps at as many as 20% of wastewater treatment plants) when they do not appear to work. The reason for this is not known. In this type of instance aluminium salts tend to be the next option, but care has to be taken about the discharge of aluminium into the receiving waters as the Environment Agency can impose limits as low as 0·15 mg Al/litre on final effluents.

There is also some concern about how efficiently anaerobic digestion will handle waste activated sludge which has been chemically dosed to remove phosphorus. Some workers suggest that there is no effect at all (e.g. Malhotra et al., 1971). Others, such as Gosset and McCarty (1978), report results which show that the use of iron-dosed sludge will affect the anaerobic digestion process adversely. Recent work, using laboratory-scale digesters, has demonstrated that changing the feed from non-iron-dosed to iron-dosed waste activated sludge caused a decrease in the gas production and the removal of volatile solids (Johnston et al., 2003).

The precise speciation of the iron which is formed in the sludge flocs when phosphate is removed in this way is not clear, but it is likely to be a

Nutrient removal

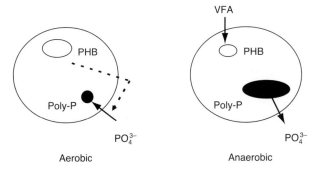

Fig. 6.6. Schematic representation of the aerobic and anaerobic reactions relating to phosphate in a BNR plant

complex hydroxy iron phosphate (de Haas et al., 2001), and a recent study has suggested that, when P is partially limiting, it will have a formula in the range of

$$Fe_4(X^{2+})PO_4(OH)_{11} \quad \text{to} \quad Fe_{42.5}(X^{2+})PO_4(OH)_{6.5}$$

Where X^{2+} is an unknown cation such as Ca^{2+} or Mg^{2+}.

However, this does not give any indication as to whether anaerobic digestion would be affected or why it might be affected.

Biological phosphorus removal

The biological removal of phosphate depends on the biochemical reactions which occur when the sludge is exposed to alternating aerobic and anaerobic regimes (Fig. 6.6). Thus, under anaerobic conditions in the absence of nitrate, bacteria will store carbon as polyhydroxybutyric acid (PHB) or as polyhydroxyalkanoate (Mino et al., 1998). Typical retention times would be two hours and the energy would be derived from the conversion of poly-phosphate to ortho-phosphate.

The process requires the presence of short-chain volatile fatty acids (acetate) for the production of the PHB. This in turn requires the presence of readily available carbon to give volatile fatty acids (VFAs). Typically, this means a BOD of more than 200 mg/litre. In the UK, sewages are likely to be too weak to operate a process for the biological removal of phosphorus successfully. This is particularly true after periods of heavy rain when the sewage is weak. This problem may be overcome by using a pre-fermentation of waste sludge to generate short-chain molecules; in other words, to use the hydrolytic/acidogenic phase of the anaerobic digestion process (see Chapter 9).

Wastewater treatment and technology

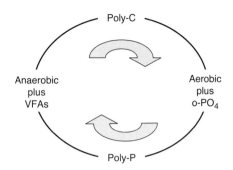

Fig. 6.7. Overall portrayal of the reactions involved in the biological uptake of phosphorus in an activated sludge plant

When these bacteria are then subjected to aerobic conditions, the PHB is broken down to provide the energy for the uptake, in excess of the metabolic requirements, of soluble ortho-phosphate, and its conversion, within the bacterial cells, to poly-phosphate. This poly-phosphate is stored in the cells (in the sludge) and is removed from the system by wastage. It must be remembered that sludge wastage will affect the sludge age and, in turn, nitrification. Therefore, plant operation should balance the needs for both. The process is one, therefore, of an alternating cycle of the formation and degradation of poly-carbon (the PHB) and poly-phosphate (Fig. 6.7). The bacteria most commonly thought of as carrying out this 'luxury uptake' of phosphorus are *Acinetobacter* spp., but, in fact, a range of different bacteria can work in this manner, possibly as a consortium (Mino *et al.*, 1998; Helmer and Kunst, 1998).

There are essentially two configurations in which these reactions can be engineered in the activated sludge process (Cooper *et al.*, 1994; Stratful *et al.*, 1999), i.e. main stream processes and side stream processes. There are several variations of the main stream option, such as the Bardenpho process, the Johannesburg process and the UCT process. The differences between them are the number of tanks used and the way in which sludges are recycled, but they all follow the basic concept of there being a sequence of anaerobic, anoxic and aerobic zones.

The layout of the Bardenpho process is shown schematically in Fig. 6.8. **Tank 1** is the main aeration tank which is designed to give a fully nitrified effluent, to oxidize BOD and to allow the uptake of ortho-phosphate. There is a high rate of recycle ($5 \times Q$) to **Tank 2**. This is an anoxic region where de-nitrification occurs and about 80%

Nutrient removal

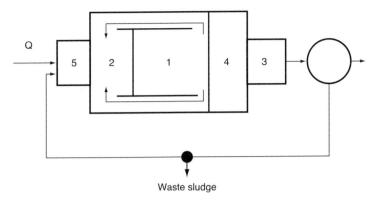

Fig. 6.8. Schematic diagram of the Bardenpho process

of the nitrates are removed. The flow, which is not recycled, passes to **Tank 3**. This is a second anoxic zone where the residual nitrates are removed. **Tank 4** is a secondary aeration unit to add oxygen to the liquors before they pass to the clarifiers. Returned activated sludge passes to **Tank 5**, which is the anaerobic stage where phosphate release occurs. Thus, the Bardenpho process is a linear process and phosphate removal is achieved through the wastage of the activated sludge.

The Phostrip process (Fig. 6.9) is a side stream process and operates with two parallel streams. One is a simple nitrifying/de-nitrifying stream using an anoxic tank before the aeration tank. The ortho-phosphate to poly-phosphate conversion occurs in the aeration tank. The ultimate removal of the phosphorus is achieved in a second stream, using a portion of the return activated sludge, and is done chemically, frequently with lime dosing, after the anaerobic breakdown of poly-phosphate and the release of ortho-phosphate. The retention time in the anaerobic reactor is about 8–12 hours, although this can be shortened by the addition of short-chain fatty acids which cause the more rapid release of the ortho-phosphate (Cooper et al., 1994).

Some of the features of the Phostrip process are that it complements existing activated sludge configurations and, as such, is easy to retro-fit with relatively low construction costs. In addition, it can work with weak sewage if the long retention time in the anaerobic stripper can be tolerated. Because the phosphorus is removed chemically it is in a form which could easily be further processed for phosphorus recovery.

Some of the performances which have been achieved by plants effecting the biological removal of phosphorus are shown in Table 6.6.

175

Wastewater treatment and technology

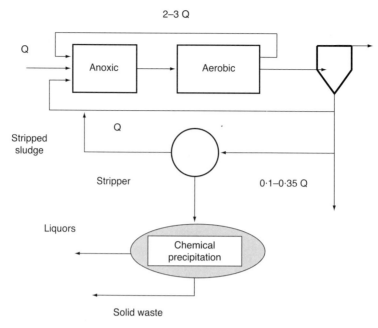

Fig. 6.9. Schematic diagram of the Phostrip process

It must be remembered that the UWWT Directive specifies limits for total phosphorus, not merely ortho-phosphate. This means that it is essential that the final settling tank performs as well as possible because the solids which it discharges will be biomass from the secondary process, activated sludge particles or humus solids. As such they will contain phosphorus. In general, biological sludges from a phosphorus-removal plant will contain a phosphorus concentration of about 4%, although, with an activated sludge having a long sludge age, the concentration could be as high as 6% (Strickland, 1998). The significance

Table 6.6. Performance of BNR plants

Process	Settled sewage		Final effluent		Source
	N (mg/litre)	P (mg/litre)	N (mg/litre)	P (mg/litre)	
3-stage Bardenpho	34	10·3	5·7	0·7	Williams, 1999
3-stage Bardenpho	22·5	8·5	6·4	0·4	Baxter-Plant, 1997
Phostrip	24·9	9·5	7·3	0·5	Baxter-Plant, 1997
Bio-P[a]	71·1	10·4	13·2	0·5	Heinzmann, 2001

[a] Bio-P is a three-tank main stream process

Nutrient removal

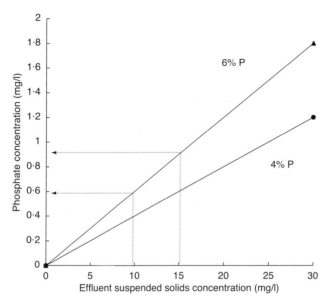

Fig. 6.10. The effect of suspended solids on the total phosphorus in the final effluent

of this can be seen from an examination of Fig. 6.10. Take, for example, an effluent containing a suspended solids concentration of 10 mg/litre. If these solids had a phosphorus content of 6%, they would contribute 0·6 mg/litre to the total phosphorus concentration of the final effluent. This would mean, for a works treating the flow from more than 100 000 pe, that the ortho-phosphate concentration would have to be less than 0·4 mg/litre. If the solids' concentration rose to 15 mg/litre, the phosphorus contribution from the solids would increase to 0·9 mg/litre. Although controlling the settlement of activated sludge is a significant problem (see Chapter 4), it has been reported that having an unaerated tank capacity of not greater than 40% of the total tank capacity should produce a sludge with good settlement characteristics.

There are disadvantages to operating a biological phosphorus removal plant. These include

- Struvite ($MgNH_4PO_4$) can be formed in the anaerobic digestion of the sludges and this can block pipes (Rabinowitz and Barnard, 1995);
- Considerable modification to an existing wastewater treatment plant is required to utilize a main stream configuration;
- During periods of high rainfall when the BOD is low, chemical dosing may be required if there is no facility for pre-fermentation. Only in a minority of cases is this not necessary (Strickland, 1998);

- The storage of waste activated sludge results in the release of phosphate into the liquors (Rabinowitz and Barnard, 1995; Strickland, 1998). This may mean that a modification of sludge handling procedures will be required. For example, anaerobic digestion of sludge would be used only after pre-anaerobic phosphate removal (Booker et al., 1999).

Phosphorus recovery

Phosphorus is the eleventh most common element on Earth. As used by modern society it must be considered as a non-renewable resource, with some 140 million tonnes of phosphate rock being mined annually throughout the world. With the impositions of the UWWT Directive for the conscious removal of phosphorus from wastewater, it would seem a logical extension to consider whether the phosphate could be recovered rather than simply removed. Chemical treatment produces a mixture of biomass and precipitated phosphate and, as such, is perhaps not the best starting point to consider recovery. Biological treatment, however, produces a phosphorus-rich biomass which will release the phosphate into solution under anaerobic conditions. These liquors have the potential to recover phosphate in a form suitable for re-use (Morse et al., 1998). The most documented approach for recovery is as calcium phosphate or struvite (magnesium ammonium phosphate). The former has received attention because the phosphate industry can use it in its production process, while struvite has excellent properties as a slow release fertilizer which poses little risk to the environment (Bridger et al., 1962; Lunt et al., 1964).

There are a number of commercial plants producing calcium phosphate (Morse et al., 1998). They use the DHV Crystalactor® process. This is based on the crystallization of the calcium phosphate onto a sand nucleus in a fluidized bed reactor (Fig. 6.11). Phosphate-rich sludge is treated in an anaerobic stripper tank to release ortho-phosphate. This process can be stimulated by the addition of acetic acid at a pH of 7·0–7·3. After settlement, a typical concentration of ortho-phosphate in the resulting liquor would be 50–80 mg/litre. This is treated with sulphuric acid to remove carbonates which would otherwise form calcium carbonate. The liquor is then fed into the fluidized bed reactor and milk of lime is added to produce calcium phosphate. The sand in the reactor has a diameter of 0·2–0·6 mm. As crystallization proceeds the particles increase in size to about 2·0 mm, and there is a periodic removal of the particles which are replaced with fresh sand.

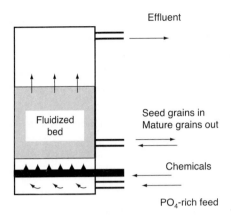

Fig. 6.11. Schematic diagram of the DHV Crystalactor^R

The vertical velocity of liquid in the bed is around 40 m/h (Gaastra et al., 1998). The particles produced by this process have a moisture content of less than 20% and are, essentially, calcium phosphate (40–50%) and sand. Their characteristics are very compatible with the requirements of the phosphate industry (Table 6.7) (Schipper et al., 2001).

A similar chemistry is used in the Kurita process which is also fully operational (Greaves et al., 1999). Calcium chloride and sodium hydroxide are mixed with the phosphate-rich stream in a fixed bed reactor packed with phosphate rock (0·5–1·0 mm diameter), and this leads to the formation of insoluble hydroxyapatite ($Ca_5(PO_4)_3OH$) which crystallizes on the packing material. As with the Crystalactor^R process, there may be a need to strip out carbonates to minimize the formation of calcium carbonate in the reactor.

The production of struvite requires the formation and crystallization of magnesium ammonium phosphate hexahydrate ($MgNH_4PO_4 \cdot 6H_2O$). In the treatment plant described by Ueno and Fujii (2001), which has a capacity of 1150 m^3/d, this is achieved by the addition of magnesium hydroxide to the liquors obtained after de-watering the waste sludge from a main stream phosphorus removal plant. The reaction uses a

Table 6.7. *Quality of the product from the Crystalactor^R process (Schipper et al., 2001)*

	P_2O_5 (g/kg ash)	Cu (mg/kg ash)	Zn (mg/kg ash)	Fe (mg/kg ash)
Industry requirements	>250	<500	<1000	<10
Crystalactor^R pellets	260	1·6	35	–

magnesium to phosphate ratio of 1 : 1 and is carried out at a pH of 8·2–8·8. This is controlled by the addition of sodium hydroxide. The reactor is a fluidized bed with a retention time of 10 days. Mixing is achieved by air injection into the bed. The pellets which are obtained under these conditions have a diameter of 0·5–1·0 mm and, after being de-watered, are sold to the fertilizer industry. The liquors from the de-watering stage, which contain fine particles of struvite, are returned to the fluidized bed reactor to act as seed material.

References

Barnes, D. and Bliss, P. J. (1983) *Biological Control of Wastewater Treatment*, E. & F. N. Spon, London.

Baxter-Plant, V. S. (1997) Personal communication.

Booker, N., Preistly, A. J. and Fraser, I. H. (1999) Struvite formation in waste water treatment plants: opportunities of nutrient recovery. *Environ. Technol.*, **20**, 777–82.

Brett, S., Guy, J., Morse, G. K. and Lester, J. N. (1997) *Phosphorus Removal and Recovery Technologies*, Selper, London.

Bridger, G. L., Salutsky, M. L. and Starostka, R. W. (1962) Metal ammonium phosphates as fertilisers, *Agric. and Food Chem.*, **10(3)**, 181–8.

Burrows, L. J. (2000) *Mathematical Modelling of the Orbal Activated Sludge Process*, PhD Thesis, University of Birmingham.

Cooper, P. F., Day, M. and Thomas, V. (1994) Process options for phosphorus and nitrogen removal from wastewater, *J. IWEM*, **8**, 84–92.

Cooper, P. F. and Wheeldon, D. H. V. (1980) Fluidized and expanded-bed reactors for wastewater treatment, *Wat. Pollut. Control.*, **79**, 286–306.

de Haas, D. W., Wetzel, M. C. and Ekama, G. A. (2001) The use of simultaneous chemical precipitation in modified activated sludge systems exhibiting biological excess phosphate removal. Part 5: Experimental periods using a ferrous-ferric chloride blend, *Water SA*, **27**, 117–34.

Downing, A. J. (1968) Factors to be considered in the design of activated sludge plant, in *Advances in Water Quality Improvement*, Gloyna, E. F. and Eckenfelder, W. W. (eds), Water Resources Symposium, University of Texas Press, pp 190–202.

Gaastra, S., Schemen, R., Bakker, P. and Bannink, M. (1998) Full scale phosphate recovery at sewage treatment plant Geestmerambacht, Holland, *Proc. Int. Conf. on Recovery of Phosphates for Recycling*, Warwick University, UK, May.

Gossett, J. M. and McCarty, P. L. (1978) Anaerobic digestion of sludge from chemical treatment, *J. Wat. Pollut. Control Fed.*, **9**, 533–42.

Greaves, J., Hobbs, P., Chadwick, D. and Haygarth, P. (1999) Prospects for the recovery of phosphorus from animal manures: A review, *Environ. Technol.*, **20**, 697–708.

Hayes, E. (1999) Personal communication.

He, H. Q., Leppard, G. G., Paige, C. R. and Snodgrass, W. J. (1995) Transmission electron microscopy of a phosphate effect on the colloid structure of iron hydroxide, *Water Res.*, **30(6)**, 1345–52.

Heinzmann, B. (2001) Phosphorus recovery in wastewater treatment plants, *2nd Int. Conf. on Recovery of Phosphorus from Sewage and Animal Wastes*, Noordwijkerhout, The Netherlands, March 12–13, CEEP, Brussels.

Helmer, C. and Kunst, S. (1998) Low temperature effects on phosphorus release and uptake by micro-organisms in EBRP plants, *Wat. Sci. Technol.*, **37(4/5)**, 531–9.

Hsu, P. H. (1975) Precipitation of phosphate from solution using aluminium salt, *Water Res.*, **9**, 115–6.

Jobbagy, A., Aimon, J. and Plotz, B. (2000) The impact of oxygen penetration on de-nitrification rates in anoxic processes, *Water Res.*, **34**, 2606–9.

Johnston, D., Carliel-Marquet, C. M. and Forster, C. F. (2003) An evaluation of the digestibility of iron-dosed waste activated sludge, *Environ. Technol.*, in press.

Kappeler, J. and Gujer, W. (1992) Estimation of kinetic parameters of heterotrophic biomass under aerobic conditions and characterisation of wastewater for activated sludge modelling, *Wat. Sci. Technol.*, **26(6)**, 125–39.

Lee, N. M. and Welander, T. (1996) The effect of different carbon sources on respirory de-nitrification in biological wastewater treatment, *J. Ferment. Bioeng.*, **82**, 277–85.

Lunt, O. R., Kofranek, A. M. and Clark, S. B. (1964) Availability of minerals from magnesium ammonium phosphates, *Agric. and Food Chem.*, **12(6)**, 497–504.

Magnusson, G., Edin, H. and Delhammar, G. (1998) Characterisation of efficient de-nitrifying bacteria strains isolated from activated sludge by 16S-rDNA analysis, *Wat. Sci. Technol.*, **38(8–9)**, 63–8.

Malhotra, S. K., Parrillo, T. P. and Hartenstein, A. G. (1971) Anaerobic digestion of sludges containing iron phosphate, *J. Sanit. Engng. Div. Proc. Am. Soc. Civ. Eng.*, 629–45.

Marais, G. v. R. (1973) The activated sludge process at long sludge ages, *Research Report No. W3*, Department of Civil Engineering, University of Cape Town.

Mino, T., Van Loosdrecht, M. C. M. and Heeijnen, J. J. (1998) Microbiology and biochemistry of the enhanced biological phosphate removal process, *Water Res.*, **32**, 3193–207.

Morse, G. K., Brett, S. W., Guy, J. A. and Lester, J. N. (1998) Review: Phosphorus removal and recovery technologies, *Sci. Total Environ.*, **212**, 69–81.

Naidoo, V., Urbain, V. and Buckley, C. A. (1998) Characterisation of wastewater and activated sludge from European municipal wastewater treatment plants using the NUR test, *Wat. Sci. Technol.*, **28(1)**, 303–10.

Oh, J. and Silverstein, J. (1999) Oxygen inhibition of activated sludge de-nitrification, *Water Res.*, **33**, 1925–37.

Orhon, D. and Ubay Çokgör, E. (1997) COD fractionation in wastewater characterisation — the state of the art, *J. Chem. Tech. Biotech.*, **68**, 283–93.

Panswad, T. and Polprucksa, P. (1998) Specific oxygen uptake rates, nitrification and de-nitrification rates of a zinc-added anoxic/oxic activated sludge process, *Wat. Sci. Technol.*, **38(1)**, 133–9.

Paul, E., Laval, M. L. and Sperandio, M. (2001) Excess sludge production and costs due to phosphorus removal, *Environ. Technol.*, **22**, 1363–71.

Rabinowitz, B. and Barnard, J. L. (1995) Sludge handling for biological nutrient removal plants, *IAWQ Yearbook*, 11–19.

Robinson, J. (2001) *Refinements to and use of a laboratory test to measure the de-nitrification potential of wastewater*, MSc Thesis, University of Birmingham.

Rössle, W. H. and Pretorious, W. A. (2001) A review of characterisation requirements for inline prefermenters: 1. Wastewater characterisation, *Water SA*, **27**, 405–12.

Schipper, W. J., Klapwijk, A., Potjer, B., Rulkens, W. H., Temmink, B. G., Kiestra, F. D. G. and Lijmbach, A. C. M. (2001) Phosphate recycling in the phosphorus industry, *Environ. Technol.*, **22**, 1337–45.

Stocks, D., Hunt, R., Connor, H., Agar, G. and Cross, R. (1994) Reconstruction of Windermere sewage-treatment works, *J. IWEM*, **8**, 10–21.

Stratful, I., Brett, S., Scrimshaw, M. B. and Lester, J. N. (1999) Biological phosphorus removal and its role in phosphorus recycling, *Environ. Technol.*, **20**, 681–5.

Strickland, J. (1998) The development and application of phosphorus removal from wastewater using biological and metal precipitation techniques, *J. CIWEM*, **12**, 30–7.

Takai, T., Hirata, A., Yamauchi, K. and Inamori, Y. (1997) Effects of temperature and volatile fatty acids on nitrification–de-nitrification activity in small-scale anaerobic–aerobic recirculation biofilm process, *Wat. Sci. Technol.*, **35(6)**, 101–8.

Tenney, M. W. and Stumm, W. (1965) Chemical flocculation of microorganisms in biological waste treatment, *J. Wat. Pollut. Control Fed.*, **37**, 1370–88.

Ueno, Y. and Fujii, M. (2001) Three years' experience of operating and selling struvite from full-scale plant, *Environ. Technol.*, **22**, 1373–81.

Upton, J. (1997) Personal communication.

Upton, J., Fergusson, A. and Savage, S. (1993) De-nitrification of wastewater: operating experiences in the US and pilot-plant studies in the UK, *J. IWEM*, **7**, 1–11.

Williams, S. (1999) Struvite precipitation in the sludge stream at Slough wastewater treatment plant and opportunities for phosphorus recovery, *Environ. Technol.*, **20**, 743–7.

7
Tertiary treatment

Introduction
The basic objective of tertiary treatment is to enhance an effluent which is already of good quality. Traditionally, tertiary treatment has been taken to refer to systems which remove solids and, because the solids are biological in nature (humus solids or activated sludge), achieve some BOD removal. With the increasingly stringent quality requirements imposed by a greater awareness for environmental protection, the term tertiary treatment tends, at the time of writing, to be applied to almost all the processes downstream of the secondary processes. Thus, it includes not only the traditional processes and their modern equivalents, but also nutrient removal (Chapter 6) and disinfection.

Solids' removal

Lagoons
These are simple low-cost processes which require little power input and maintenance and, as well as being considered as a tertiary treatment prior to discharge to river, they can also be thought of as a pre-treatment which reduces the cost of disinfection by ozone or UV radiation. Basically, they give a period of additional settlement, but with retention times of about 50 hours. Ideally, a series of small lagoons is preferable to a single large one. While their objective is to remove solids (Fig. 7.1), because they are shallow, typically about 1 m deep, they may improve the bacteriological quality because of light

Wastewater treatment and technology

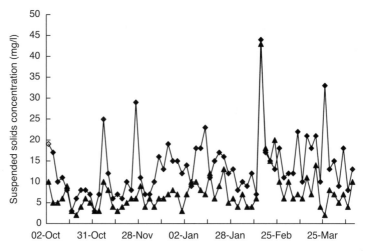

Fig. 7.1. Influent (♦) and effluent (▲) quality from a tertiary treatment lagoon

(UV) penetration. Sunlight can also result in the growth of algae in the summer months and a scum-board should be used to protect the final weir. Typical loading rates and performances are given in Tables 7.1 and 7.2. They do need to be cleaned every one or two years to remove the solids which have accumulated during this time. The extended hydraulic retention time also provides a degree of buffering against occasions when the effluent quality is less than adequate. However, they take up a lot of room and are only a rural technology. They are a hazard to children and must be fenced to account for this.

Grass plots

These are very simple to operate if there is land available. Perhaps up to three hectares may be needed. They require a distribution channel, a

Table 7.1. Summary of the characteristics of tertiary treatment processes

Process	ΔSS (%)	ΔBOD (%)	Loading
Lagoons	45–75	30–70	$0.5 \text{ m}^3/\text{m}^2$ per day
Grass plots	50–70	50–75	$0.2–0.5 \text{ m}^3/\text{m}^2$ per day
Reed beds	75–80	75–90	$0.104–0.208 \text{ m}^3/\text{m}^2$ per day
Microstrainers	30–80	25–75	$300–700 \text{ m}^3/\text{m}^2$ per day
Rapid gravity filters	60–85	c. 70	$10–12.5 \text{ m}^3/\text{m}^2$ per hour[a]

[a] Maximum load

Table 7.2. Performance of lagoons

Works	Number of lagoons	Total area (m^2)	HRT (d)	Loading (m^3/m^2 per day)	BOD (mg/litre)		SS (mg/litre)	
					In	Out	In	Out
I	2	8 × 10^4	12	0·46	27	14	21·8	11·7
II	2	9110	2·2	0·77	25	9	23·6	12·4
III	3	225	1·9	0·53	31	14	15	8·3

collection channel and a slope of 1 in 60 to 1 in 100. The soil structure and vegetation remove the solids. There will also be some removal of BOD and ammoniacal-nitrogen. There will be a solids' build-up and this means that plots have to be rested and, in some cases, renovated. Grass plots can require an appreciable input of manpower if performance is to be maintained. The plots need to be rotated, grass needs to be cut and the channels kept clean. A typical arrangement has three plots so that one can be in operation, one can be drying out and resting to permit the soil bacteria to degrade the accumulated solids and the third can be renovated or have the grass cut. Renovation is, perhaps, an extreme measure as it involves removal of the top layer of soil, a re-grading of the surface soil and re-seeding, a process which may take as long as 12 months. Nevertheless, it should be considered at periods of every five years.

Typical loadings and performance are given in Tables 7.1 and 7.3, but it must be recognized that periods of prolonged heavy rain may result in solids being washed out into the discharge, which would detract from the normal performance. Although, at first sight, the concept of grass plots is a simplistic one, they do have their problems and at least one UK water company has reached the conclusion that the use of new

Table 7.3. Performance of grass plots

Works	Loading (m^3/m^2 per day)[a]	BOD (mg/litre)		SS (mg/litre)	
		In	Out	In	Out
I	0·09	16·6	4·7	27·5	12·5
II	0·57	9·7	2·6	21	4
III	0·37	19·4	12·5	21	13

[a] Based on the area in use

grass plots is not to be recommended. Certainly, subsurface flow reed beds are smaller, have greater reliability and can be considered as a clear alternative to grass plots.

Upflow clarifiers

These use pebble beds, wedge wire or plastic mesh in the final settlement tank to give an enhanced removal of the solids. The beds are 150 mm deep and, if pebbles were being used, they are 6–10 mm diameter. The beds are located 150–300 mm below top water level. There is some removal by physical straining, but there is also a significant element of flocculation within the solid bed which assists in the removal of solids. Typically, an efficiency of 30–60% removal of suspended solids can be expected. The head loss across the clarifier bed is small, about 25 mm. This means that such systems can be fitted to existing settlement tanks (Table 7.4 and Fig. 7.2). However, if this is done, there will be a loss of settlement capacity during the cleaning phase and cleaning must be done to remove the solids which have been retained within the bed. Clarifiers can be designed as purpose built systems and they are designed on the basis of an upflow velocity of 1 m/h at maximum flow. When purpose-built tanks are being designed, some workers recommend that the number of tanks should be such that there is one more than is necessary to allow for beds to be taken out of operation for cleaning.

Cleaning, which involves lowering the water level to below the bottom of the bed and washing the medium, will be needed every two to three days for wedge wire or plastic mesh. If pea gravel is used, the cleaning frequency can be extended to a weekly sequence. This necessity for cleaning means that the use of upflow clarifiers can require a regular input of manpower.

Table 7.4. Types of upflow clarifiers

Works	Type of media	Installed as	Mean hydraulic loading rate (m^3/m^2 per day)
I	Pebble bed	Separate clarifier	2·20
II	Wedge wire	Integral unit	5·67
III	Wedge wire	Integral unit	8·56
IV	Plastic mesh	Separate clarifier	7·46
V	Plastic mesh	Integral unit	10·4

Tertiary treatment

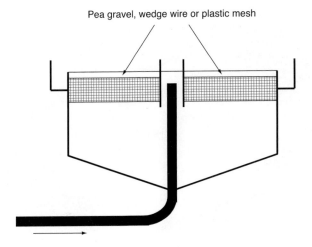

Fig. 7.2. Schematic diagram of an upflow clarifier

Microstrainers

A microstrainer consists of a rotating cylindrical drum, the walls of which are made up of a series of fine mesh windows, usually made of stainless steel, supported on a steel framework (Figs 7.3 and 7.4). The effluent from the secondary treatment process enters at one end and filters outwards through the mesh. As the drum rotates the particles which have been retained by the mesh are washed off by jets of water, at a pressure of 100–350 kPa, using 2–5% of the filtrate. Typically, microstrainers have diameters of 1–3 m and are 0·3–0·45 m in length. The mesh diameters are 15–60 μm and the peripheral speed is 0·5–0·7 m/s. One of the advantages of using microstrainers is the low head loss (50–150 mm). Typical performance and loading

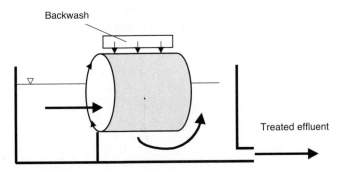

Fig. 7.3. Schematic diagram of a microstrainer

Fig. 7.4. *A microstrainer being used for tertiary treatment*

data are given in Table 7.1. Although the washing sequence can also incorporate irradiation with ultraviolet light, bacterial growth can start to block the mesh and cause the performance to deteriorate over time. This means that microstrainers need to be taken out of operation so that the mesh fabric can be thoroughly cleaned with hypochlorite. This may be needed every three to four months.

Rapid gravity filters

These operate in the same way as those used for the treatment of potable water, i.e. the secondary effluent is passed down through a bed of sand (about 1·2 m deep) with the driving force being the head of water above the sand. Typically, this is about 1·5 m. Thus, their role is the removal of fine suspended solids which have not been removed at the secondary settlement stage. Although, at first sight, the process is a simple one, the ways in which the solids are removed are complex. Some of the forces responsible for particle entrapment are

- interception   direct contact along a streamline with a bed grain;
- inertia   particle deflected from its streamline into a bed grain;
- gravity   settlement following Stoke's law;
- diffusion   random transport of particles.

In rapid gravity filters there is deep penetration of the bed by the solids and as these solids accumulate there is an increase in the head loss across the bed. This means that the bed has to be cleaned and this is

Tertiary treatment

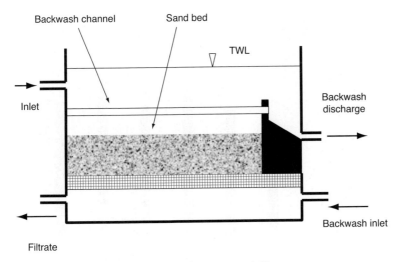

Fig. 7.5. *Schematic diagram of a rapid gravity sand filter*

achieved by backwashing using air scour, water scour or a combination of the two. If water is used, it would be treated water and, typically, about 5–10% of the throughput is used. For the combined option, water is used at a rate of $11–36\,m^3/m^2$ per hour, together with air scour at a rate of $55–110\,m^3/m^2$ per hour. Typical sand diameters are 0·8–1·7 mm and the sand is laid on top of a bed of gravel (Fig. 7.5). The treated water is discharged into the underdrains by nozzles set in a suspended floor. These nozzles are also used to introduce the back-wash fluids into the bed.

There are a number of operational aspects which need consideration, specifically

- reconciling the conflicting requirements of throughput and head loss;
- loading;
- backwashing.

The question of throughput and head loss can be dealt with in one of three ways. The filter can be operated with increasing head, with a declining rate or with a flow control system. In the increasing head mode, a constant output is achieved, but the head gradually increases over the filter cycle. In the declining rate mode, the head is kept constant, but this means that the throughput decreases over the filter cycle. With a flow control system the outlet is fitted with a valve which, initially, is only slightly open. As the solids accumulate in the

Wastewater treatment and technology

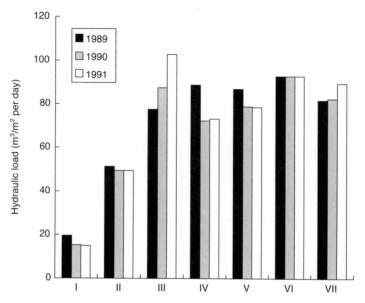

Fig. 7.6. Hydraulic loading rates for seven rapid gravity filters (Lander, 1994)

bed, this valve opens gradually to maintain a constant flow rate. In his survey of rapid gravity filters operated by Severn Trent Water, Lander (1994) reported that 75% were operated in the increasing head mode. The actual length of the cycle would be determined by head loss, time or total flow.

Figures 7.6 and 7.7 show the mean annual hydraulic and suspended solids' loadings applied to seven rapid gravity filters over a three-year period. These need to be compared with the normally accepted maximum of $10-12.5 \, m^3/m^2$ per day at maximum flow. A suspended solids' removal efficiency of 60–85% would normally be expected (Fig. 7.8 and Table 7.1). Certainly the review by Lander (1994) showed that, with one exception, the mean suspended solids' concentration in the effluents from seven rapid gravity filters was appreciably less than 10 mg/litre.

As with most processes, there can be operational problems. Sand can be lost from filters if the design is inadequate or if the distribution of the backwash water and air causes high turbulence. High flows generated by storm conditions can result in high solids' concentrations being discharged from the secondary treatment processes and this, in turn, leads to shorter cycle times. Another problem which can plague rapid gravity filters is the phenomenon of 'mudballing'. This is the formation of a coating of organic matter around the filter medium. This is often

Tertiary treatment

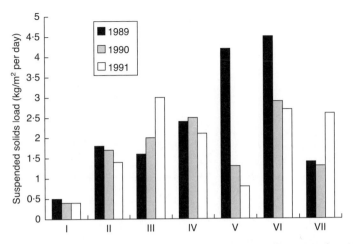

Fig. 7.7. Suspended solids load applied to seven rapid gravity filters (Lander, 1994)

the result of sludge being carried over from the secondary stage, coupled with the inability of the backwash regime to remove it effectively. The effect of this biofilm on the sand is to reduce the overall density of the particle. For example, a particle with a dry matter content of 80% would have a relative density of 2.04 g/ml, whereas one with a dry matter content of 25% would have a relative density of 0·89 g/ml (Lander, 1994). This means that unless the flow rate of the backwash water is reduced significantly there will be a severe loss of the filter medium.

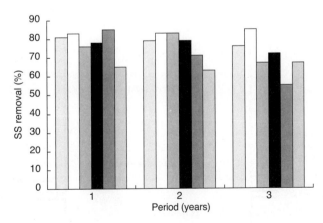

Fig. 7.8. The removal of suspended solids by six rapid gravity sand filters over a three-year period

191

Table 7.5. A comparison of the main characteristics of rapid gravity and deep bed filters

	Rapid gravity bed	Tetra deep bed
Sand size (mm)	1·0	2·0–3·0
Sand depth (m)	0·8	1·8
Gravel depth (m)	0·3	0·46
Surface loading (m^3/m^2 per hour)	5–10	10–20
Solids removal (%)	60–85	60–85
Backwash volume (%)	5–10	2–5

Deep bed filters

The main differences between deep bed filters and rapid gravity filters are given in Table 7.5. The most significant difference, as the name suggests, is the depth of sand used in the filter. A second important aspect is that the Tetra filter (Fig. 7.9) uses a 'T' block arrangement to support the bed rather than a suspended floor. Also there is an absence of nozzles in the deep bed. One of the advantages of the deep bed is that it gives a greater penetration of solids which, in turn, permits longer run times and higher rates of filtration.

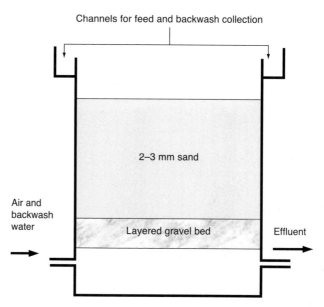

Fig. 7.9. Schematic diagram of a Tetra deep bed filter

Upflow sand filters: the Dynasand filter

The Dynasand filter is a moving bed system with a counter current flow of water and sand (Fig. 7.10). The water to be treated is introduced through a distributor in the lower section of the unit into sand which is dirty. As it passes upwards through the bed it encounters sand which is progressively cleaner and is moving downwards. The dirty sand at the base of the unit is conveyed by an air-lift pump to a sand washing unit at the top of the filter. The abrasive nature of the air-lift separates the dirt particles from the sand, so that, in the washing unit, only a small upward flow of water is needed to achieve the

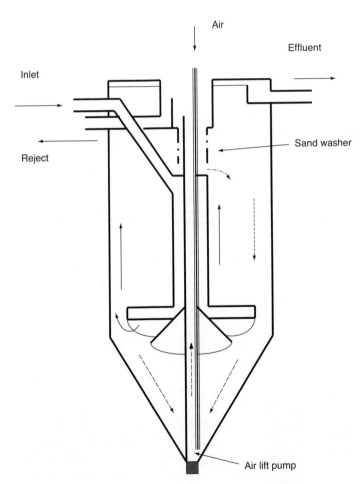

Figure 7.10. Schematic diagram of a Dynasand filter showing the downflow of sand (– – – –) and the upflow of water (———)

separation of the dirt from the heavier sand particles which sink back into the bed. The wash water containing the dirt is then discharged through the wash water outlet and returned to the head of the treatment works. Typically, the wash water would be 8–10% of the flow being treated. The Dynasand filter is, therefore, a continuous process which has no need for a separate backwashing stage. It is sold commercially in a range of sizes with surface areas of up to $5\,m^2$ and is installed as multiple units where necessary. The sand which is used is 1–2 mm diameter and design surface loading rates, at average flow, are $8.5\,m^3/m^2$ per hour. The performance expressed as a removal percentage will depend on the concentration of suspended solids in the water being treated but, in most cases, a final effluent containing less than 10 mg/litre would be expected.

Reed beds

Although reed beds (Figs 7.11 and 7.12) were originally designed for the treatment of wastewater and sludges (Chapter 5), they are increasingly considered as tertiary treatment processes (Cooper, 2001). Under these conditions, they are fed with a fully nitrified secondary effluent. They also have a role in the treatment of storm sewage (Griffin and Upton, 1999). This can be done either in separate storm reed beds or as combined storm and tertiary treatment beds (Green et al., 1999). The beds use 5–10 mm pea gravel as their matrix and are, essentially, the same as those described in Chapter 5. For tertiary or combined treatment, an area of $1\,m^2$/capita is used, and this will produce an effluent with a BOD of less than 5 mg/litre and a total suspended solids concentration of less than 10 mg/litre (Cooper, 2001). However, at least one water company in the UK uses a value of $0.7\,m^2$/pe for tertiary treatment (Green and Upton, 1995). For storm treatment alone, the

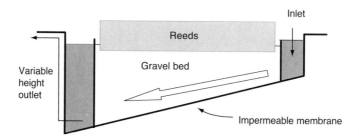

Fig. 7.11. Schematic diagram of a reed bed

Tertiary treatment

Fig. 7.12. An operational reed bed in conjunction with a rotating biological contactor (courtesy of Copa Ltd)

beds are sized at $0.5\,\text{m}^2/\text{pe}$ (Green *et al.*, 1999). The performance of reed beds as tertiary treatment units is given in Table 7.6.

Disinfection

The Bathing Water Directive (76/160/EEC) requires that waters designated as bathing areas should conform to specified microbiological and physico-chemical standards. These bathing areas may be in coastal or inland waters. Arguably, the microbiological standards (Table 7.7) are the main concern for engineers designing or operating wastewater treatment plants. Even a high-quality effluent will contain bacteria

Table 7.6. Quality of effluents from reed beds being used for tertiary treatment

Works		Influent	Effluent
I	BOD (mg/litre)	11·6	2·7
	SS (mg/litre)	25·5	5·7
	Amm.N (mg/litre)	3·4	0·7
II	BOD (mg/litre)	9·1	1·0
	SS (mg/litre)	18·4	4·5
	Amm.N (mg/litre)	6·6	1·9
III	BOD (mg/litre)	13·2	3·6
	SS (mg/litre)	20·2	4·1
	Amm.N (mg/litre)	2·8	3·7

Table 7.7. *Microbiological standards required by the Bathing Water Directive*

Parameter	Guide value (90 percentile)	Mandatory value (95 percentile)	Min. sampling frequency
Total coliforms/100 ml	500	10 000	Fortnightly[a]
Faecal coliforms/100 ml	100	2000	Fortnightly[a]
Faecal streptococci/100 ml	100	–	b
Salmonella/litre	–	0	b
Enteroviruses/10 litre	–	0	b

[a] May be reduced if the biological quality is good
[b] Checked when inspection suggests that a deterioration in water quality may have occurred or that the organism might be present

and viruses. Therefore, where a discharge may have the potential for affecting a bathing area, disinfection may be required. The objective of a disinfection scheme is to kill any microbes (bacteria, viruses) in the water. The extent with which bathing areas in the UK comply with the mandatory standards specified by the Directive is shown in Fig. 7.13 for the period 1995–2000. Over the same time period, there was approximately a 50% compliance with the guideline values.

Chlorine

Probably the most widely used disinfectant is chlorine, which can be added either as a gas, Cl_2, or as solid sodium hypochlorite, $NaOCl$.

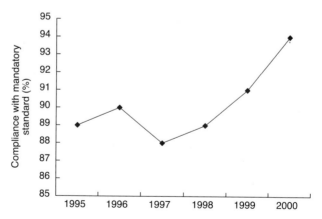

Fig. 7.13. *UK compliance with the mandatory standards specified by the Bathing Waters Directive*

Tertiary treatment

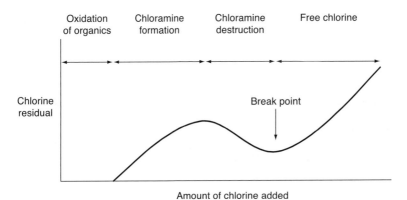

Fig. 7.14. *Reactions of chlorine, water and ammoniacal-nitrogen*

Chlorine will react with any ammoniacal-nitrogen and organic matter in the sewage and will also react with the water itself:

$Cl_2 + H_2O = HOCl + HCl$

$HOCl \leftrightarrow H^+ + OCl^-$

$Cl_2 + NH_3 = HCl + NH_2Cl$

$Cl_2 + NH_2Cl = HCl + NHCl_2$

Gaseous chlorine, hypochlorous acid and hypochlorite ions are classified as 'free chlorine', whereas the chloramines are classified as 'bound chlorine'. Thus, as chlorine is added to water, the proportions of the two types of chlorine will alter depending on the degree of contamination of the water by ammonia and organics. The point at which free chlorine starts to be available is the break point (Fig. 7.14). Free chlorine inactivates bacteria by disrupting the integrity of the cell membrane. It will also damage the nucleic acids (Haas and Engelbrecht, 1980). Chloramines, although they utilize some of the added chlorine, are also capable of acting as disinfectants, albeit more slowly than free chlorine (Shih and Lederberg, 1974). Chloramines are always likely to be formed when sewage is treated with chlorine, unless complete nitrification has occurred in the secondary treatment stage, and they can be toxic to fish and invertebrates.

The killing rate for chlorine is traditionally given by

$$\frac{dN}{dt} = -kN \qquad (7.1)$$

where k is the rate constant. This is certainly correct for clean water. However, for dirty water it may be more correct to use

$$\frac{dN}{dt} = -kNt \qquad (7.2)$$

In both cases, N is the number of bacteria and t is the contact time in minutes.

Bacterial death is also a function of the concentration of chlorine (C in mg/litre), the temperature and the pH, and for sewage an empirical relationship has been developed:

$$N_t = N_o(1 + 0 \cdot 23C \cdot t)^{-3} \qquad (7.3)$$

where N_o is the initial number of bacteria. On site, chlorine can be used as gas stored in cylinders, as solid sodium hypochlorite or it can be generated by the electrolysis of sodium chloride. The control of chlorination is important and it is usual to add a fixed dose with a fixed contact time. The dose will depend on the type of treated sewage which is being chlorinated, but typical values are 5–10 mg/litre for settled sewage and 2–8 mg/litre if secondary treatment is used. When chlorination is being used in conjunction with a long sea outfall, it is not uncommon to use the time in the outfall pipe as the contact time.

The problem with the use of chlorine is that chlorinated compounds are released into the environment. The formation of trihalomethanes (THMs) is of particular concern. They are produced by the reaction between organic compounds and chlorine and are carcinogenic. The best known is chloroform, $CHCl_3$. There are legal limits for THM concentrations in drinking water and, therefore, if the receiving water is to be used for abstraction, the appropriateness of chlorine should be carefully considered. Thus, from an environmental point of view, there is a 'chlorine debate'. Should it be used? Certainly, if a new chlorine usage is being considered, it is important to establish a baseline for the ecology of the receiving waters before chlorine is used, so that any damage, or lack of it, can be properly quantified.

Ozone

Ozone (O_3) is an alternative to chlorine, but it has to be produced on site by applying an electrical discharge (5–25 kV) to dry air (Fig. 7.15). This will produce a gas which contains a relatively small proportion (1–2% V/V) of ozone. Ozone is a powerful oxidant but it is not very

Tertiary treatment

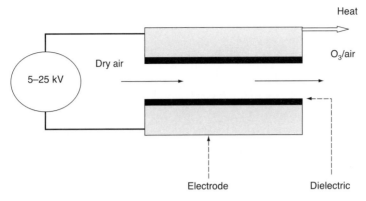

Fig. 7.15. Schematic diagram of an ozone generator

soluble in water. It can also be broken down by heat. It kills microorganisms by generating free radicals in solution. It affects the enzymes and nucleic acid of bacteria and damages the nucleic acid core of viruses (Ishizaki *et al.*, 1987). The product of concentration and time, as with chlorine, describes its effectiveness as a disinfectant. The values required to achieve a 99% kill of *Escherichia coli* are in the range of 0·001–0·2 (Hall and Sobsey, 1993). However, the different species which are of concern to the engineers operating wastewater treatment plants have an appreciable range of sensitivities/resistance. For example, polio virus is more resistant than *E. coli.* which in turn is more resistant than *Salmonella typhimurium* (Fig. 7.16) (Farooq and Akhlaque, 1983). The presence of solids can reduce the impact of

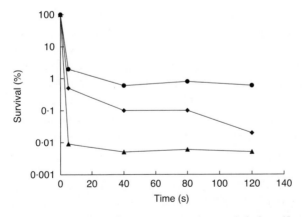

Fig. 7.16. Microbial inactivation by ozone in an activated sludge effluent (Farooq and Akhlaque, 1983); poliovirus 1 (●), E. coli (◆) and S. typhimurium (▲)

199

Wastewater treatment and technology

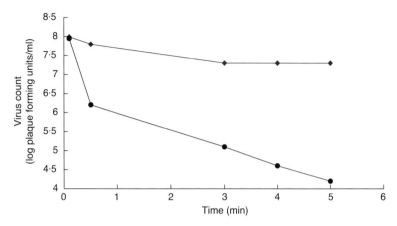

Fig. 7.17. *Effect of solids on the effect of ozone on poliovirus (Kaneko, 1989); no suspended solids (●), activated sludge — 10 mg/litre (◆)*

ozone, as is shown in Fig. 7.17. An examination of the use of ozone to disinfect a secondary treated effluent, prior to discharge to estuarine waters, showed that a 2 log reduction in E. coli was readily achieved with ozone concentrations of 11–21 mg/litre and a contact time of 30 minutes (Eastman et al., 1993). Although it is a relatively expensive option, ozone has the advantage that it does not give any residues or by-products which can be disseminated into the environment.

Care needs to be taken with the materials of construction at an ozonolysis unit because it is such a powerful oxidant, and a disinfection facility using ozone should be fully alarmed for the safety of the operating personnel. This would be based on the Control of Substances Hazardous to Health Regulations and the Occupational Exposure Standards for ozone.

Ultraviolet radiation
Ultraviolet (UV) radiation will also kill bacteria. The critical wavelength is 253·7 nm which damages the cell's DNA. There are expressions which may be used to describe the effectiveness of microbial inactivation by UV (Severin, 1980):

$$\frac{N}{N_o} = e^{-KPt} \tag{7.4}$$

where N_o and N = the initial and final numbers of micro-organisms per ml after time t, P = the intensity of light reaching the target organisms ($\mu W/cm^2$), K = the rate constant. However, it is doubtful whether this could be used to quantify the efficiency of the treatment of sewage effluents, which will contain a range of species with a range of susceptibilities. The extent of the efficiency of UV, all other things being equal, does depend on the species being targeted. In general, the order of resistance to UV is protozoan cysts > bacterial spores > viruses > vegetative bacteria (Chang et al., 1985).

Ultraviolet disinfection systems are based on mercury lamps in quartz tubes. The lamps are submerged at a controlled liquid depth in an open channel and the typical exposure time is 6–10 seconds. This contact time needs to be compared with 5–10 minutes for ozone and 15–30 minutes for chlorine. Andreadakis et al. (1999) have reported that to achieve a faecal coliform count of less than 2000/100 ml, a dose of 30–60 mW s/cm^2 was needed for secondary effluent, with lower doses (about 10 mW s/cm^2) being needed for tertiary effluents. This confirmed the earlier work by Job et al. (1995), who demonstrated that a dose of 60 mW s/cm^2 would achieve better than 3·5 logarithmic reduction in total thermotolerant coliforms.

Until recently, it had been considered that the use of UV would be restricted to the disinfection of very clean water which would not restrict light penetration. Certainly, the effluent needs to be of good quality with a low concentration of suspended solids and, certainly, there is a need to ensure that the lamps are kept clean. This is particularly true after activated sludge systems, where iron salts have been added to remove phosphates because fine, semi-colloidal particles of iron hydroxide can quickly coat the tubes sufficiently to reduce the passage of radiation and diminish the effectiveness of the process (Gehr and Wright, 1998). As a disinfectant, it has the advantage of not producing residuals or by-products and, therefore, can be thought of as being environmentally friendly.

Most of the UK water companies with coastlines are examining the potential of UV treatment (Lowe, 1999; Langley et al., 1999), and there have been several installations in the UK of UV for the disinfection of treated sewage. The Trojan ultraviolet technology, which is quite widely used in North America, has been installed at a number of sewage works in the UK. It is an open channel design (Jackson, 1994) and can use low or high intensity lamps, depending on which process is being used. The use of high intensity lamps reduces the number of lamps which are required and, therefore, the space requirements. The UV4000^{TM} system which has been installed at, for example, Swansea with a peak flow of 1300 litre/s, and Weston-super-Mare with a peak

Table 7.8. UV treatment of final sewage effluents

Parameter				
Peak flow (m³/d)	4160	5680	6430	51 000
Secondary process	Oxidation ditch	Activated sludge	Trickling filter	Activated sludge
Effluent quality SS:BOD	30:30	20:20	30:30	30:12
Transmittance (%)	65	65	65	65
Dose (W/s per cm²)	30 000	35 000	30 000	30 000
Faecal coliforms/100 ml	37	57	80	175

Table 7.9. Effect of UV treatment on bacterial numbers in secondary effluent (Perrot and Baron, 1995)

	Counts/100 ml	
	Before treatment	After treatment
E. coli		
Mean	$6·55 \times 10^4$	43
Maximum	$5·49 \times 10^5$	1770
Minimum	$4·76 \times 10^3$	9
Faecal streptococci		
Mean	$2·34 \times 10^4$	42
Maximum	$2·31 \times 10^5$	1350
Minimum	$1·20 \times 10^3$	1·4

flow of 1000 litre/s, can be fitted with an electronic modulating system which controls the output from the lamps as the effluent quality and flow vary. The performance of UV treatment is shown in Tables 7.8 and 7.9.

References

Andreadakis, A., Mamais, D., Christoulas, D. and Kabylafka, S. (1999) Ultraviolet disinfection of secondary and tertiary effluent in the Mediterranean region, *Wat. Sci. Technol.*, **40(4–5)**, 253–60.

Chang, J. C. H., Ossof, S. F., Lobe, D. C., Dorfman, M. H., Dumais, C. M., Qualls, R. G. and Johnson, J. D. (1985) UV inactivation of pathogenic and indicator micro-organisms, *Appl. Environ. Microbiol.*, **49**, 1361–5.

Cooper, P. (2001) Constructed wetlands and reed-beds: Mature technology for the treatment of wastewaters from small populations, *J. CIWEM*, **15**, 79–85.

Eastman, G. M., Harker, R. J. and Ibbetson, C. C. (1993) Advanced oxygen treatment of coastal municipal sewage, *Wat. Sci. Technol.*, **27(5–6)**, 425–37.

Farooq, S. and Akhlaque, S. (1983) Comparative response of mixed cultures of bacteria and viruses to ozonation, *Water Res.*, **17**, 809–12.

Gehr, R. and Wright, H. (1998) UV disinfection of wastewater coagulated with ferric chloride: recalcitrance and fouling problems, *Wat. Sci. Technol.*, **38(3)**, 15–23.

Green, M. B., Martin, J. R. and Griffin, P. (1999) Treatment of combined sewer overflows at small wastewater treatment works by constructed reed-beds, *Wat. Sci. Technol.*, **40(3)**, 357–64.

Green, M. B. and Upton, J. (1995) Constructed reed bed: Appropriate technology for small communities, *Wat. Sci. Technol.*, **32(3)**, 339–48.

Griffin, P. and Upton, J. (1999) Constructed wetlands: A strategy for sustainable wastewater treatment at small treatment works, *J. CIWEM*, **13**, 441–6.

Haas, C. N. and Engelbrecht, R. S. (1980) Physiological alterations of vegetative micro-organisms resulting from chlorination, *J. Wat. Pollut. Control Fed.*, **53**, 1976–89.

Hall, R. M. and Sobsey, M. D. (1993) Inactivation of hepatitis A virus (HAV) and MS-2 by ozone and ozone-hydrogen peroxide in buffered water, *Wat. Sci. Technol.*, **27**, 371–8.

Ishizaki, K., Sawadaishi, K., Miura, K. and Shinriki, N. (1987) Effect of ozone on plasmid DNA of *Escherichia coli in situ*, *Water Res.*, **21**, 823–7.

Jackson, G. F. (1994) Operational experience of ultra-violet disinfection of sewage effluent in Jersey, *Europ. Wat. Pollut. Control*, **4**, 18–21.

Job, G. D., Trengove, R. and Realey, G. J. (1995) Trials using a mobile ultraviolet disinfection system in South West Water, *J. CIWEM*, **9**, 257–63.

Kaneko, M. (1989) Effect of suspended solids on the inactivation of poliovirus and T2-phage by ozone, *Wat. Sci. Technol.*, **21**, 215–9.

Lander, J. (1994) Wastewater rapid-gravity filtration in Severn Trent Water, *J. IWEM*, **8**, 256–68.

Langley, P., Stringer, R. and Lang, G. (1999) Yorkshire Water's Coastcare, *Proc. Inst. Civ. Eng. — Civ. Engng.*, **132**, 59–64.

Lowe, N. R. (1999) Green Sea — achieving Blue Flags in Wales, *Proc. Inst. Civ. Eng. — Civ. Engng.*, **132**, 53–8.

Perrot, J. Y. and Baron, J. (1995) The disinfection of municipal wastewater by ultraviolet light: A French study, *Wat. Sci. Technol.*, **32(7)**, 167–74.

Severin, B. F. (1980) Disinfection of municipal wastewater effluents with ultraviolet light, *J. Wat. Pollut. Control Fed.*, **52**, 2007–18.

Shih, J. L. and Lederberg, J. (1974) Effect of chloramine on *Bacillus subtilis* deoxyribonucleic acid, *J. Bacteriol.*, **125**, 934–45.

8
Sludge handling and disposal

Introduction

Sewage sludge is the term used for the solid component of sewage, separated from the liquid (which becomes the treated effluent) during the treatment process. The sludge comes from both the primary and secondary treatment stages. Typically, sludges contain 95–99% water, but have a significantly higher solids' concentration than the raw sewage (99·97% water). The sludges generated during the treatment of municipal sewage require disposal. As can be seen from Table 8.1 the amounts can be considerable. In the UK the amount of sludge currently being produced is about $1 \cdot 5 \times 10^6$ tonnes dry solids per year (Hudson, 1995; Davis, 1996; Bruce and Evans, 2002). Between 2 and 2·5 litres of organic residue, at approximately 3% dry solids, are produced by every person served by a wastewater facility. The full implementation of the Urban Wastewater Treatment

Table 8.1. *Predicted sludge disposal routes by 2005 for some European countries (Bruce and Evans, 2002)*

Country	Agriculture (%)	Landfill (%)	Incineration (%)	Other (%)	Total (1000 t DS/a)
France	66	0	34	0	1172
Germany	50	18	30	2	2787
Portugal	30	60	0	10	359
Netherlands	28	17	49	0	401
Greece	8	92	0	0	99
UK	71	7	21	1	1583

Sludge handling and disposal

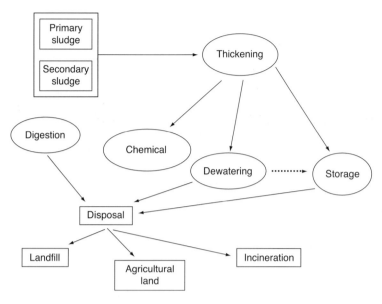

Fig. 8.1. Sludge handling and disposal options

Directive has had two effects in relation to sludge disposal. It has forbidden disposal to sea and it has specified that all discharges from sewage treatment works treating flows from population equivalents of more than 2000, including those which discharge to coastal and estuarine waters, should receive an element of treatment. Together, this means that more sludge will require land-based disposal routes. This additional amount has been estimated as being 50–60% of the current quantity (Hudson, 1995; Davis, 1996). The options which are now available for disposal are incineration, land-fill or application to agricultural land, while the handling/treatment options are reduced to de-watering, anaerobic digestion or chemical treatment (Fig. 8.1). Raw sewage will contain pathogens, from humans and from some of the industrial inputs, for example, abattoirs and hospitals (Table 8.2). Many of them can be ignored as they cannot survive in the sewerage system at temperatures well below that of their human and animal hosts. However, there are some which have the potential to cause harm and many of them will become bound within the sludge flocs and concentrate in the sludge.

There is considerable concern throughout the world about the safety of land-based disposal routes, particularly to agricultural land. In the UK, the key elements which affect the disposal of sludge to agricultural land are

205

Table 8.2. Common pathogens found in sewage sludge (Lewis-Jones and Winkler, 1991; Strauch, 1998)

Bacteria	Viruses	Parasites
Salmonella spp.	Poliovirus	**Protozoa**
Shigella spp.	Coxsackievirus A and B	Entamoeba histolytica
Escherichia coli	Echovirus	Giardia lamblia
Pseudomonas	Adenovirus	Toxoplasma gondii
Clostridium perfringens	Hepatitis A	Sarcoystis spp.
Clostridium botulinum	Rotavirus	Cryptosporidium parva
Bacillus anthracis		
Listeria monocytogenes		**Nematodes**
Leptospira spp.		Ascaris lumbricoides
Vibrio cholerae		Ancylostoma duodenale
Campylobacter spp.		Toxocara canis
Staphylococcus		Toxocara cati
Streptococcus		Trichuris trichiura
		Cestodes
		Taeina saginata
		Taeina solium
		Echinococcus granulosus

- The EC Directive on the use of sewage sludge in agriculture, adopted in the UK as The Sludge (Use in Agriculture) Regulation, Statutory Instrument Number 1263 (S.I. 1263) in 1989. Later, this was amended to S.I. 880 which implemented EU Directive (86/278/EEC). Currently, this code is being revised and the industry awaits an overhaul of the guidelines.
- The Environmental Protection Act 1990 (EPA 1990). Part II of the Act affects sludge and waste disposal and places the responsibility (duty of care) of waste disposal on those who generated the initial waste. Integrated Pollution Prevention Control (IPPC) was proposed to replace Part I of the EPA 1990. This was initiated and brought major waste processing facilities within the jurisdiction of the IPPC legislation, and included sewage treatment plants which exceeded 50 tonnes per day. Consequently, plants which fall into this category must have an IPPC authorization to operate and must comply with specified emissions standards.
- The Environment Act 1995. This places sludge processing and disposal under direct regulation and monitoring by the Environment Agency (EA). The EA are the official regulator of waste management practice, including sludge to land.

Table 8.3. *International biostandards for sludge*

Country	Biostandard (dry solids basis)		
	Faecal indicator	Salmonella	Parasitic ova
France	3 cfu (MPN) enterovirus/10 g	8 MPN/10 g	3 helminth ova/10 g
Italy	–	1000/g	
Switzerland[a]	100 enterobacter/g	–	zero parasitic ova/g
Norway	1000 faecal coliforms/g	3/g	–
USA[b]	1000 faecal coliforms (MPN)/g	3/4 g	

[a] For land growing fodder crops or vegetables
[b] Standard for Class A biosolids and is based on either the faecal indicator or the salmonella counts
cfu = colony forming units

- Code of Practice for Agricultural Use of Sewage Sludge, 2nd edition (1996). This provides guidance regarding the safe use of sewage sludge in agriculture.

In the USA, the US Environment Agency has defined two types of sludge: Class A and Class B. These categories are defined in Sections 405 (d) and (e) of the Clean Water Acts with the definitions being based on the numbers of pathogens or indicator species present in the sludge (Bryant, 1993). The more rigorously treated Class A solids, which must contain fewer than 1000 MPN (most probable number) faecal coliforms per g of total solids, are permitted the wider use. The USEPA 503 Regulations define the time-temperature criteria for obtaining Class A solids:

$$D = \frac{50 \cdot 05 \times 10^6}{10^{0 \cdot 14T}} \qquad (8.1)$$

where D = contact time (days), T = temperature (°C). This equation assumes that the process is operated in a batch mode or with true plug-flow, that the minimum temperature is 50 °C and that the solids' concentration of the sludge is less than 7%.

The Swiss standard is more stringent, requiring a count of fewer than 100 enterobacteria per g. Examples of other biological standards are given in Table 8.3. These values may well be modified by the new European Directive on sludge, which is currently being drafted by the Waste Management Unit of the Environment Directorate (DG 12) of the European Commission.

Table 8.4. The requirements for 'treated' and 'enhanced treated' sludge

	E. coli reduction	E. coli numbers	Salmonella spp
Treated	99% or 2 log$_{10}$	10^5/g dry solids. Up to 31-12-03 as a 90 percentile basis. From 1-1-04 as a maximum	not applicable
Enhanced treated	99·9999% or 6 log$_{10}$	<10^3/g dry solids sampled monthly	zero/2 g dry solids

Within the UK, the House of Commons' Environment Committee stated that by 2002 all sewage sludge being disposed of to land should be stabilized and pasteurized. Concern about the safety of food has led to an agreement between the British Retail Consortium, representing the food retailers, and Water UK (an association of the water companies), which affects the application of sewage sludge to agricultural land (Anon, 1999). This is known as the ADAS 'Safe Sludge Matrix'. It has had a number of impacts

- It has phased out the use of raw or untreated sludge so that, after the end of December 2001, the application of all such sludges to any agricultural land was banned;
- It has restricted the use of sewage sludge on agricultural land unless the sludge has been subjected to advanced treatment (Tables 8.4 and 8.5).

On this basis, treated sludge is classified as one in which there has been a 99% (2 log$_{10}$) reduction in the indicator organism *Escherichia coli* and its fermentability has been significantly reduced. Advanced treatment is one which achieves a 6 log$_{10}$ reduction in the numbers of *E. coli*. In

Table 8.5. Simplified form of the safe sludge matrix

Crop	Untreated	Sludge digested	Advanced treatment
Fruit, salad	No	No	Yes[a]
Vegetables	No	No[a]	Yes[a]
Horticulture	No	No	Yes[a]
Combinable and animal food crops	end 31 Dec 1998[a]	Yes[a]	Yes
Grass for silage	No	Yes[a]	Yes
Grass for grazing	No	Yes[a]	Yes
Maize silage	No	Yes[a]	Yes

[a] Various provisos apply

Fig. 8.2. A picket fence thickener

addition, the sludge should contain fewer than 1000 cfu/100 g dry solids. This means that, where advanced treatment is necessary for sludge disposal, something analogous to pasteurization will be required so that harmful bacteria will be killed. Pasteurization can be achieved by radiation (Swinwood and Wilson, 1990) or by heat. Thermal pasteurizing conditions have been defined as 70 °C/30 minutes or 55 °C/four hours (Department of the Environment, 1996). The conditions for achieving this will be discussed later.

Sludge thickening

Sludges are normally thickened prior to any type of treatment. Indeed, one economic analysis has considered a pre-thickening/digestion/post-thickening sequence and shown that, compared with the other options involving digestion, it gave the lowest overall unit cost (Bruce et al., 1990).

Gravity thickening

The simplest way of achieving the initial thickening is to use a picket fence thickener (Fig. 8.2), which can be operated as a batch or a continuous process. This is really no more than an extended period of extra settlement, with the phase separation being aided by slow stirring by a 'picket fence', typically at a peripheral speed of 1–3 m/min (Price and Alder, 1985; Hoyland et al., 1989). This has the effect of

Table 8.6. *General criteria for sizing continuous consolidation tanks* (Hoyland et al., 1989)

Type of sludge	Plan area (m^2 day/t SS)	Solids concentration (kg SS/m^3)
Primary	9	60–90
Primary + humus	13	50–80
Primary + activated	20	45–70
Humus	60	30–45

- producing channelling in the sludge which aids the upward movement of water;
- releasing any gas bubbles;
- preventing 'bridging' by the solids.

The design criteria for this process operated in the continuous mode have been described by Hoyland et al. (1989) and will vary with the type of sludge being processed (Table 8.6).

Centrifugation

Centrifugation is an alternative option for thickening sludges (Davarinejad and Voutchkov, 1994; Tolan et al., 1989). One philosophy suggests that waste activated sludge should be kept separate and not co-settled in the primary tanks. This gives a better separation of the primary solids. At one works operated by Severn Trent Water these are thickened by centrifugation after treatment with polymers. Part of the stream is centrifuged to give about 18% solids and then mixed with untreated sludge to give a final concentration of 6%, the concentration specified for anaerobic digestion. Waste activated sludge by itself can also be thickened by centrifugation. At one site an Alpha Laval decanting centrifuge was used in conjunction with a cationic polymer to condition the sludge. This gave a centrifugal force of up to 2000 × gravitational acceleration, depending on the rotational speed. A typical set of performance data is shown in Fig. 8.3. The mean polyelectrolyte usage was 4 kg/tonne of dry solids (DS). However, work by Davarinejad and Voutchkov (1994) has shown that when mixtures of primary and secondary sludges were centrifuged, the composition of the mixtures had an appreciable effect on the performance, with the cake concentration increasing with increasing amounts of primary sludge in the mix.

Sludge handling and disposal

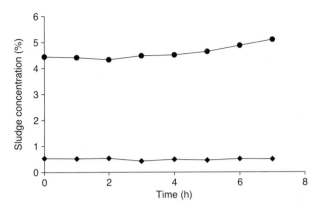

Fig. 8.3. Variation in the centrifuge feed (♦) and product (●) sludge concentration (start time = 0845)

Belt thickeners

A gravity belt thickener (Fig. 8.4) may also be used. This consists of an endless moving belt with a weave which allows the passage of water. Sludge, which has usually been conditioned with a polyelectrolyte, is fed onto the belt and the water drains out under gravity as the sludge moves along the belt. At the discharge end of the bed there is further de-watering by the compression caused as the sludge turns over on itself. Again the performance varies with the type of sludge being processed and whether a sludge conditioner has been used. For waste activated sludge treated with a polymer conditioner, a concentration of 6·0% can be achieved (Westoll, 1999). Similar performances have been reported by Thomas *et al.* (2000) — 6·5% for surplus activated sludge; and by Farrell *et al.* (2001) — 5·5% for the excess sludge from a submerged biological contactor.

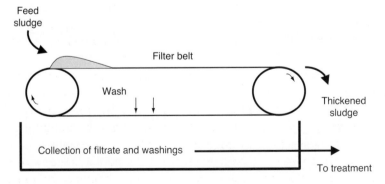

Fig. 8.4. Schematic diagram of a belt thickener

211

Wastewater treatment and technology

Dissolved air flotation

Although dissolved air flotation (DAF) has been widely used in water treatment, both for solids removal and sludge thickening, it is not a popular option in the UK for thickening sewage sludge (Hudson and Lowe, 1996). This is not the case in other parts of the world. The basic concept of the process is that air bubbles are used to lift the sludge to the surface of a tank where a sludge cake is formed. This can then be removed. As is shown in Fig. 8.5, the components of a DAF unit are

- a contact tank where chemicals are added to impart an element of hydrophobicity to enhance the attachment of the sludge particles to the air bubbles;
- a saturator where air is added under pressure to a proportion of the treated water;
- a flotation tank. This can be thought of as having two zones: a reaction zone where the bubbles and the sludge attach and a flotation zone where the solids are taken to the surface and form a cake.

The chemicals which are used are those used throughout the water industry for coagulation and flocculation, namely iron salts, aluminium salts and organic polymers. There are two types of saturator, packed and unpacked, but there seems to be no specific pressure used for achieving saturation. The reported values range from 200 to 800 kPa. The other design parameters are also quite variable. For example, the air to solids ratio can vary from 0·01 to 0·1 and the surface hydraulic loading rate can be between 20 and 330 m^3/m^2 per day. The performances reported

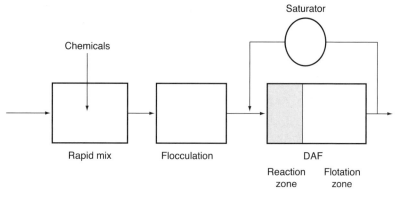

Fig. 8.5. Schematic diagram of a DAF plant

Sludge handling and disposal

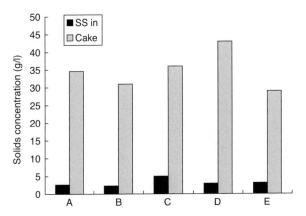

Fig. 8.6. *Performance of DAF plants thickening waste activated sludge (Haarhoff and Bezuidenhout, 1999)*

for full-scale DAF plants used for thickening sewage sludges show a wide scatter. A recent report has examined the operation of five DAF plants which were thickening waste activated sludge without the use of coagulents, and Fig. 8.6 shows the degree of thickening which was achieved (Haarhoff and Bezuidenhout, 1999).

Sludge rheology

The rheological characteristics of sewage sludges tend to become increasingly non-Newtonian as their suspended solids' concentrations increase. Specifically, they become thixotropic and exhibit yield stresses. It is important, therefore, that the relationships between the rheological properties and the suspended solids' concentrations are known, and that sludges are not thickened to concentrations at which yield stress values give problems with pumping, mixing or heat transfer (Moeller and Torres, 1997; Brade et al., 2000). For primary sludges the critical concentration is about 12%.

There are many types of viscometer which may be used to measure the rheological characteristics of sludges. Some are field units, others are strictly laboratory tools. One field unit, which was developed specifically to examine rheology in relation to head loss, is shown in Fig. 8.7. The sludge level in the reservoir was measured against time and, by applying Equations (8.2) and (8.3), the basic rheological constants for Equation (8.4) can be derived

$$\tau_W = \frac{D\Delta P}{4L} \tag{8.2}$$

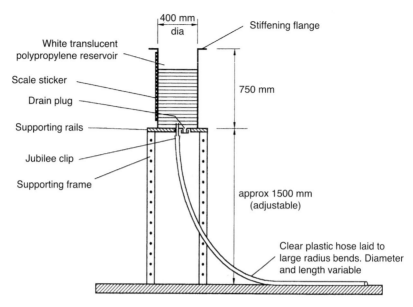

Fig. 8.7. Tube viscometer for determining the rheological properties of sludge

$$\Delta P = \rho_S g h \qquad (8.3)$$

$$\tau_W = \tau_o + K\left(\frac{8V}{D}\right) \qquad (8.4)$$

where τ_W = wall shear stress (N/m²), D = tube diameter (m), L = tube length (m), ρ_S = sludge density (kg/m³), h = head (m), V = velocity of flow (m/s), K = constant. For primary sludges, the values of the yield stress, τ_o, and K have been shown (Papacrivopoulos, 1985) to vary with the suspended solids' concentration (S, %)

$$\tau_o = 4\cdot353S - 14\cdot08 \qquad (8.5)$$

$$K = 0\cdot0736S - 1\cdot233 \qquad (8.6)$$

The published literature on sludge rheology using laboratory units is based on a variety of viscometers. A comparison of the results obtained from the two viscometers which were available in the Public Health Laboratories at Birmingham is given in Fig. 8.8. There was a very strong correlation between the viscosity measured by a multi-speed Fann viscometer with that measured by a single-speed Contraves viscometer. The relationship was significant at the 99% level. This means that although more work of this type is necessary, it should be

Sludge handling and disposal

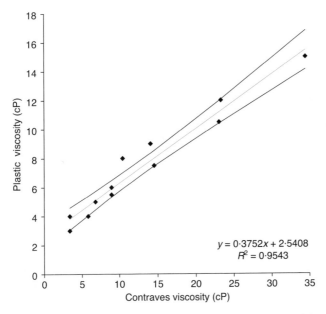

Fig. 8.8. A comparison of the activated sludge viscosities measured by different viscometers

possible to compare the relative values of measurements made with different viscometers.

The variations in the yield stress of activated sludge as the suspended solids' concentration was increased are given in Fig. 8.9 (Forster, 2002). The data show that activated sludge exhibited an appreciable yield stress at solids' concentrations above 9–11 g/litre. The relationships shown in Fig. 8.9 are comparable with those reported previously for activated sludge (Dick and Ewing, 1967), and the existence of a critical solids' concentration for a yield stress to be exhibited has also been reported by Christensen *et al.* (1993). Figure 8.9 also shows that iron dosing, for the removal of phosphate (see Chapter 6), resulted in sludges having a lower yield stress than those before iron dosing. The reduction in the yield stress was assumed to be due to the flocculation of the sludge particles by the ferrous/ferric ions. A similar effect can be seen when the sludge had been dosed with a cationic polymer before thickening.

In the event of viscometers not being available, an approximation to the rheological characteristics of a sludge, based on the sludge type, its solids concentration and the temperature, can be obtained from the WRc's *Technical Report TR185* (Frost, 1983).

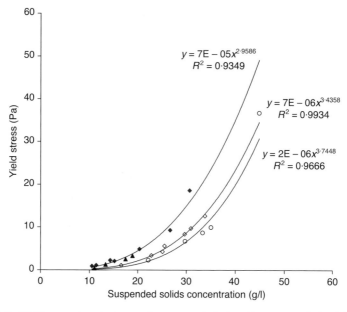

Fig. 8.9. Yield stress in relation to the activated sludge (AS) suspended solids concentration showing the regression lines

Anaerobic digestion

Essential elements
Anaerobic digestion is used to stabilize sludges by degrading the putrescible components. The resultant sludge, although not totally odour-free, does not usually generate any further odours on storage. The degradation of these putrescible components is achieved by a sequence of integrated biochemical processes (Fig. 8.10).

The hydrolytic bacteria initiate the process and are responsible for the degradation of long chain compounds, such as proteins, carbohydrates and lipids, using a range of exo-enzymes, such as proteases, cellulases and lipases. There is a wide diversity of bacterial species in this group, which is a mixture of both facultative (can tolerate oxygen) and obligate (oxygen is toxic) anaerobic bacteria, and populations of 10^{10}–10^{11} per gram of volatile solids would be expected in a mesophilic sludge. Where there is a significant amount of polymeric material in the substrate, as in the case with sludge, this stage is thought to be the rate limiting step (Speece, 1983).

The acidogenic stage is undertaken by a large group of bacteria, some of which may also be part of the hydrolytic group. In general, they are a

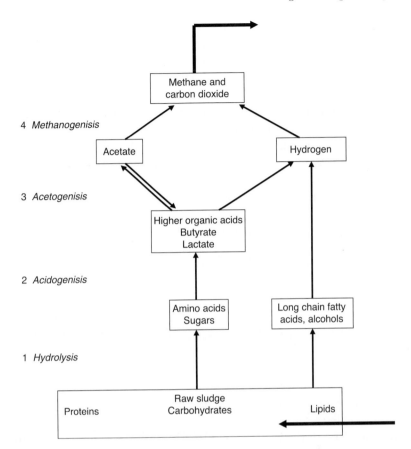

Fig. 8.10. Schematic representation of the anaerobic digestion process

relatively robust group with a rapid doubling time, about 20 minutes, so that populations of 10^{11}–10^{12} per gram of sludge could be expected. Their role is to convert the products of the hydrolytic step to carboxylic acids. Thus

$$C_6H_{12}O_6 + 2H_2O \rightarrow 2CH_3COOH + 4H_2$$
$$C_6H_{12}O_6 + 2H_2 \rightarrow 2CH_3CH_2COOH + 2CO_2 + 2H_2O$$
$$C_6H_{12}O_6 \rightarrow CH_3CH_2CH_2COOH + 2CO_2 + 2H_2$$

The main reaction would be to produce acetate and hydrogen and, generally, the hydrogen would be consumed as it was produced by methanogens, homoacetogens and sulphate reducing bacteria. However, in the event of shock loading, excessive amounts of both acetate and hydrogen could be produced, leading to a drop in pH and, ultimately, a failure of the digester by 'souring'. Feedback control loops within the acidogens would then lead to the utilization of alternative pathways which would produce propionate and butyrate. Hydrogen plays a critical role in determining the products of acidogenesis and this is why the partial pressure of hydrogen has a role as a control parameter, and why the relative concentration of acetate, propionate and butyrate can be used as an assessment of digester stability.

There are two main groups of acetogenic bacteria: the obligate hydrogen-producing acetogens and the homoacetogens. The homoacetogens produce acetic acid and other carboxylic acids from carbon dioxide. The former group, which is comprised of only two types, *Syntrobacter* and *Syntrophomonas*, convert the carboxylic acids and other short chain molecules to acetic acid, hydrogen and carbon dioxide

$$CH_3CH_2COOH + 2H_2O \rightarrow CH_3COOH + CO_2 + 3H_2$$

$$CH_3CH_2CH_2COOH + 2H_2O \rightarrow 2CH_3COOH + 2H_2$$

These reactions are not thermodynamically favourable and will only proceed if the partial pressure of hydrogen is low (10^{-4}–10^{-5} atmospheres).

The methanogens are obligate anaerobes and will only utilize a limited range of substrates: acetate, formate, methanol, methylamine, carbon monoxide and carbon dioxide. There are two main types. The acetoclastic bacteria split acetate into methane and carbon dioxide. About 70–75% of the methane produced in a digester comes from this route. Some of the carbon dioxide which is formed will be bound as bicarbonate and increase the alkalinity of the sludge. Hydrogenotrophic bacteria produce the rest of the methane by the reduction of carbon dioxide by dihydrogen. The acetoclastic bacteria are slow growing species. Typically, they would have mean generation times of 2–3 days, compared with the acidogenic bacteria which double in 2–3 hours.

Design and operation

The essential features of the operation of an anaerobic digester are a pH of 7·0–7·5, a very definite anaerobic environment and, in most cases, a mesophilic temperature (30–37 °C). In addition, the digester contents must be well mixed to avoid the accumulation of grit/silt at the bottom of the digester and to prevent stratification where localized 'pockets' of acidity could build-up. Mixing also enhances gas release and reduces the risk of scum formation. It can be achieved mechanically or by gas re-circulation. With the latter, gas is taken from the headspace above the sludge, compressed and injected at pre-selected locations within the sludge using either injection lances, diffusers or gas lift systems (WEF, 1992). This approach has the advantage of there being no moving parts within the main body of the digester and is the most common method with modern designs. Mechanical mixing systems include pumped re-circulation, draft tube mixers and turbine mixers. The power input for mixing is generally taken as being between 5 and 8 W/m^3 (WEF, 1992).

A heat input is required to maintain the mesophilic (or higher) temperature. It must bring the feed sludge up to the digestion temperature and it must compensate for any heat losses. External heat exchangers, using hot water as the heat source, are probably the most common method. The main requirements for heat exchangers are that they should have large sludge pathways to minimize the risk of blockages, that the thermal resistance of the wall and boundary layer should be as low as possible and that the maximum temperature on the water side should not be greater than about 70 °C to avoid baking on the sludge side. Biogas can be burnt to produce this heat. It has a calorific value of about 22–25 MJ/m^3. However, it is wise to make provision for an auxiliary fuel source for periods of low biogas production or high heating requirements. Consideration should certainly be given to using combined heat and power (CHP). That is to say, providing hot water to heat the digesters and generating electricity for use on site or even for export (Walker, 1996). Submerged combustion is an alternative heating option (Williams *et al.*, 1993). Although this was first examined in the 1970s and can achieve pasteurization (Kidson and Ray, 1984; Holt and Hudson, 1999), it has not, at 2003, been used in the UK other than for testing purposes.

Anaerobic digesters are constructed with an aspect ratio (sidewall depth:diameter) of 0·9–1·2 and a floor slope of about 12° (Fig. 8.11). The roof is either floating or fixed (Fig. 8.12), but the latter will require a separate storage facility. The essential feature of the roof, of whichever type, is that it prevents air entering and forming a potentially explosive mixture in the head space.

Digesters are usually designed to give a specific hydraulic retention time or organic loading rate and it has been suggested that, when analysing existing digesters, these should be based on the effective volume to allow for space lost by the accumulation of grit (Brade and Noone, 1981). The derivation of the effective volume will necessitate the use of tracers such as lithium chloride, bromine 82 (for the liquid

Fig. 8.11. Anaerobic digesters (courtesy of Severn Trent Water)

phase) and gold 198 (for the solid phase). As can be seen from Table 8.7, the values which have been used for the hydraulic retention times and the loading rates do vary, but it has been suggested that there is an upper limit of 3·2 kg volatile solids/m^3 per day governed by the rate of accumulation of toxic materials (WEF, 1992). The day-by-day performance can be judged by

- the biogas production — it should not vary significantly;
- the methane content of the gas — it should be about 60–66% and it should not vary significantly;
- the volatile fatty acid (VFA) concentration — it should be 200–500 mg/litre;
- the alkalinity — it should be 3500–5000 mg/litre;
- the pH — it should be 7·0–7·5.

The US design manual (Zickefoose and Hayes, 1976) for anaerobic digestion also recommends that for a stable digestion

- the VFA:alkalinity ratio should be $< 0\cdot4$;
- the propionate:acetate ratio should be $< 1\cdot4$;
- the acetate concentration should be < 800 mg/litre;
- the butyrate concentration should be < 15 mg/litre.

With a well-stabilized digestion a reduction in volatile solids of about 50% could be expected. This would be accompanied by a gas

Sludge handling and disposal

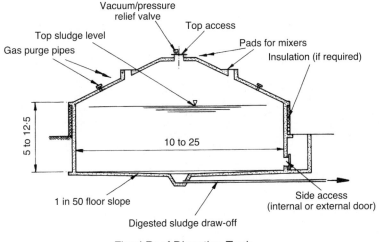

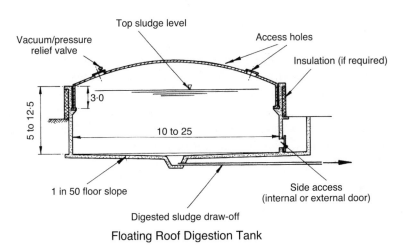

Fig. 8.12. *Fixed roof and floating roof digestion tanks*

production of about 1 m³/kg volatile matter destroyed (Table 8.7). The performance of digesters can be disrupted by components of the sludge (Table 8.8). Detergents will always be present in sewage and will become adsorbed onto the sludge. Cationic and non-ionic detergents will have little effect on the digestion process, but if the concentration of anionic detergents (measured as Manoxol OPT) is greater than 1·5%, on a dry solids basis, inhibition will occur. Heavy metals, particularly cadmium, chromium, copper, nickel and zinc, will also have an effect on the digestion process. Their effects are synergistic and Mosey

Table 8.7. *Typical performance data for mesophilic anaerobic digesters*

Loading (kg VS/m^3 d)	1·8	1·01–2·04	1·63–1·97	2·3[a]
Retention time (d)	18·6	13·6–24·8	18·4–23·6	15[a]
VS reduction (%)	40·6	40·0–49·4	42·1–44·2	48·9
Gas yield (m^3/kg VS destroyed)	0·8	1·03–1·2	0·39–0·60	0·77
Source	Andrew and Salt, 1987	Brade et al., 1982	Brade et al., 1982	Brade and Noone, 1981

[a] Based on the effective volume

(1976) has developed an expression, based on their equivalent weights being additive, to quantify these effects

$$K = \frac{(Zn/32·7) + (Ni/29·4) + (Pb/103·6) + (Cd/56·2) + (Cu/47·4)}{\text{sludge solids concentration} \quad (\text{kg/litre})} \quad (8.7)$$

The concentrations of the metals in the sewage are expressed as mg/litre so that K has the units of meq/kg. It has been shown that, if K exceeds a value of 400 meq/kg, there will be a 50% chance of digester failure. Therefore, efforts should be made to maintain K at a value of less than 170 meq/kg to ensure that there is a 90% probability that the performance of the digester will not be affected.

The methanogens require a redox potential of −200 to −420 mV and if nitrates are present the redox potential in the digester can be moved in a more positive direction, thus disrupting the overall performance. However, the increasing use of nitrification and de-nitrification (Chapter 6) means that this particular problem should be diminishing.

Table 8.8. *Concentrations of toxic material causing significant (20%) inhibition of anaerobic digestion*

Toxin	Concentration (mg/kg dry solids)
Trichlorethane	20
Carbon tetrachloride	200
Nickel	2000
Copper	2700
Zinc	3400
Anionic detergent	1·5–2·0%

Sludge handling and disposal

(a) (b)

Fig. 8.13. Egg-shaped digesters: (a) during construction (courtesy of Atkins Water) and (b) in operation

Digesters can also be built with an egg-shaped configuration (Fig. 8.13). The advantage claimed for this configuration is that its shape creates a circulation pattern (Fig. 8.14), which is better for keeping solids in suspension. Also, the reactor has a lower surface area at the top and bottom, therefore minimizing the potential for solids accumulation. This is borne out by the study by Li *et al.* (1996) who reported that there was no noticeable scum after 10 years of continuous operation. Egg-shaped digesters are common in mainland Europe and they have also been installed in other parts of the world, for example, at the Deer Island wastewater treatment plant at Boston, USA and at the Woodman Point facility in Western Australia. The digesters at Deer Island are 27·3 m in diameter and 40 m in height. Traditionally, this type of configuration has not been used in the UK; however, they are now starting to be installed. Thames Water, for example, have specified this design for the new works at Reading, as have United Utilities at their Southport plant.

Dual digestion

As well as stabilizing the sludge microbiologically, any digestion process must optimize the destruction of volatile solids and the production of

Wastewater treatment and technology

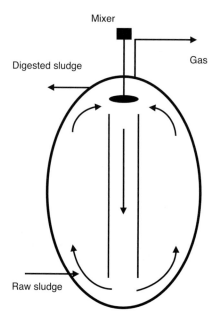

Fig. 8.14. Mixing pattern in an egg-shaped digester

gas in relation to the size of the digester. One way of achieving this process optimization is to separate the acidogenic stage from the more demanding methanogenic stage. This means that both stages can be optimized. Hanaki *et al.* (1987) have reported that, even with a complex substrate, two-stage systems gave a quicker overall treatment. Furthermore, it was possible to waste sludge from the acidogenic unit without disrupting the methanogenic stage (Cohen *et al.*, 1982). Other advantages of two-stage systems are a faster start-up time and the prevention of inhibition (Tanaka and Matsuo, 1986; Komatsu *et al.*, 1991). Coupling these facts with the need for a greater microbial stabilization has led to the concept of dual digestion, which is a two-stage system, with one stage being operated in the thermophilic range and one in the mesophilic range. For example, Oles *et al.* (1997) have used an acidogenic first-stage in the thermophilic range, followed by a mesophilic, methanogenic stage, and have demonstrated a significant reduction in the time required to achieve a specified degradation of organic matter, compared with a single-stage digester. Zhao and Kugel have reported similar results (1996). However, both these systems used retention times of more than one day. Indeed, Oles *et al.* (1997) recommend a retention time of 2–3 days in the thermophilic reactor, but it is not clear whether shorter retention times and,

Sludge handling and disposal

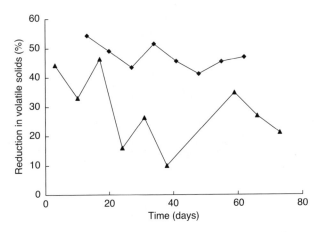

Fig. 8.15. A comparison of the overall volatile solids reduction by a single-stage mesophilic digester (▲) and a dual thermophilic/mesophilic digestion system (♦) (Roberts et al., 1999)

therefore, smaller and less costly reactors, would be as effective. Work at the laboratory-scale would suggest that shorter retention times in the thermophilic reactor could be used (Roberts *et al.*, 1999). Figure 8.15 compares the performance of a single-stage mesophilic digester having a hydraulic retention time of 20 days with that of a dual digestion system in which the retention times in the thermophilic and mesophilic digesters were four hours and 12 days respectively.

Enhancing anaerobic digestion
Currently, there is considerable interest in enhancing the performance of anaerobic digesters; in other words, either processing more sludge or producing more biogas in existing assets. As was shown in Fig. 8.10, any anaerobic digestion process involves a series of biochemical reactions: hydrolysis, acid production and, finally, the formation of methane. Hydrolysis is the rate limiting stage when feed-stocks with high suspended solids are being used (Eastman and Ferguson, 1981). This is also true when significant proportions of surplus activated sludge are being digested. This is due to the poor digestibility of the bacterial cells and the extracellular polymeric material which is an integral part of the sludge. A number of physical and chemical processes which effect a degree of solubilization of solids have, therefore, been examined as a pre-treatment stage prior to the main anaerobic digestion. These include (van Lier *et al.*, 2001)

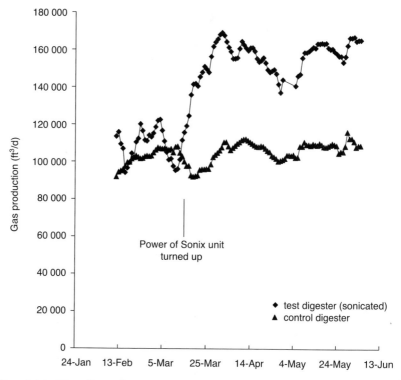

Fig. 8.16. The effect of sonication on anaerobic digestion (courtesy of Atkins Water)

- mechanical methods;
- ultrasonic treatment;
- chemical treatment;
- thermal hydrolysis;
- enzymatic hydrolysis.

The use of ultrasound to break bacterial cells is well known. Sonication has also been used to disrupt activated sludges and, as such, has been used to enhance anaerobic digestion. Clark and Nujjoo (2000) have shown that, with laboratory-scale chemostats, sonication can induce an increase in specific methane yield of over 60% with a hydraulic retention time of 15 days. However, work by Quarmby et al. (1999) has shown that, although the use of freshly sonicated sludge did give improved gas yields, the effect was negated if there was storage of the sonicated sludge. This problem can be overcome by sonicating the sludge as it is fed to the digester. Figure 8.16 shows the results of a trial with two identical 3785 m³ (1 000 000 US gallons) digesters which were operated with a 20-day hydraulic retention time. The feed was a mixture of primary sludge (40%) and thickened waste activated sludge (60%). The feed to the test digester was treated with a Sonix^{TM} 5 horn installation which had a running duty of 10 kW. The Sonix^{TM} unit was turned on at low power in early February and turned up to its running power in March. As can be seen from Fig. 8.16, the effect, particularly after about two weeks, was very marked. In making comparisons of the results from sonication trials it must be recognized that there does not appear to be a consistent view about the dosage which is required or how to describe what dose or what solids' concentration should be used.

Thermophilic digestion
The anaerobic digestion process can also be operated in the thermophilic (55–60 °C) range, but, traditionally, this has not been a particularly popular option in the UK. One of the reasons for this has been that thermophilic digestion has been deemed to be potentially unstable (Kiyohara et al., 2000). However, the introduction of Class A solids and the concept of enhanced treatment means that its use should be re-examined. The work reported by Aitken and Mullennix (1992) certainly suggests that thermophilic anaerobic digestion is a viable process for achieving the Class A criteria. The use of the higher temperature suggests that the reactions by which the putrescible fraction was degraded takes place more quickly. The growth rate of *Methanosaeta* is known to be about 2·5 times greater than in the mesophilic range. It has also been shown that the optimal hydraulic retention time in a thermophilic anaerobic digester is 10 days (Alatiqi et al., 1998). The use of thermophilic anaerobic digestion is being actively pursued in Denmark as a means of increasing the capacity of existing assets (Nielsen and Petersen, 2000).

Enzymatic hydrolysis
This is essentially a dual digestion process in which the first stage is optimized for maximum hydrolysis and pathogen reduction rather than being optimized for maximum acid production. At the moment, results are only available from laboratory scale digesters, but a patent application is pending and a full scale installation is nearing completion. Thus, the first stage treats the sludge at a temperature of 42 °C for two days, and the second stage is operated as a conventional mesophilic (35 °C) digester with a hydraulic retention time of 20 days. In a 12 month study, using a sludge containing $2·4 \times 10^6$ E. coli/g dry solids, the results

showed that the sludge from the enzymatic hydrolyser/mesophilic digestion process contained an average of 3315 E. coli/g dry solids, whereas mesophilic digestion alone produced a sludge containing $1·2 \times 10^5$ E. coli/g dry solids (Mayhew et al., 2001).

Enhanced treatment options

Aerobic digestion

An alternative way of producing a sludge which conforms to the Class A or enhanced treatment criteria is to use thermophilic treatment in the aerobic mode. There are two variations of this type of process: thermophilic aerobic digestion (TAD) and autothermal thermophilic aerobic digestion (ATAD). Both processes operate in the 55–65 °C temperature range. The essential features of the TAD process are a source of heat, good insulation, an efficient aeration system and heat exchangers to transfer heat from the treated sludge to the incoming raw sludge (Fig. 8.17). The heat for the ATAD process is derived from the metabolic activity of the bacteria in the sludge as they degrade the putrescible fraction. As organic matter is converted into new cells, free energy is released. In an aeration tank, such as that used in the activated sludge process, this energy is dissipated into the environment but in a purpose-built, well-insulated tank, such as is used in the ATAD process, it accumulates, resulting in a rise in temperature. This effect

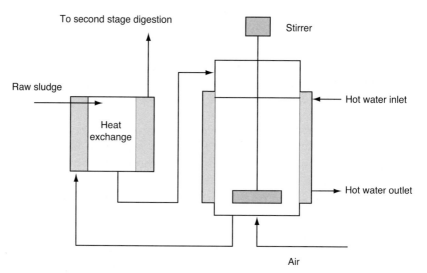

Fig. 8.17. Schematic diagram of a TAD plant

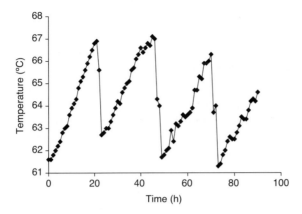

Fig. 8.18. *Temperature cycles in an ATAD reactor*

can be seen in Fig. 8.18, which shows the temperature profile in a batch fed ATAD tank. The total energy released (ΔF in kJ/litre) can be calculated from

$$\Delta F = 14 \cdot 6 E (COD_{initial} - COD_{final}) \qquad (8.8)$$

where E is the efficiency of heat retention.

Since the heat released (kJ/litre) is approximately equal to the change in temperature and since, in most cases, E is about 70% (Jewell and Kabrick, 1980), the temperature rise can be expressed as

$$\Delta T = 2 \cdot 4 (COD_{initial} - COD_{final}) \qquad (8.9)$$

The essential criteria for ATAD are

- good insulation;
- good oxygen transfer — the efficiency should be better than 10%;
- sufficient biodegradable organics in the raw sludge — the generation of 100–150 kJ/litre is considered appropriate.

Currently, there are only a few TAD and ATAD manufacturing companies around the world, the main differences between them being the type of aeration system and the heat exchangers.

The handling characteristics of TAD and ATAD sludges, which means their pumping properties and how readily they will filter, also need to be considered. Recent work has shown that both types of sludge have poor filtration characteristics, with capillary suction time (CST) values (see later) of 1250 and 1500 s respectively (Riley and Forster, 2001). This has also been noted previously (Cohen, 1977; Jewell and Kabrick, 1980). The reason for this would appear to be related to the very small size of the sludge particles after digestion. The ATAD sludge had a mean particle diameter of 53 µm and the TAD sludge had one of 45 µm (Riley and Forster, 2001). The pumping properties of the sludges can be judged in terms of their yield stress. Both types of sludge had yield stress values which were significantly lower than activated sludge at the same concentration. At a concentration of 40 g SS/litre, TAD sludge had a yield stress of 1·32 Pa and ATAD sludge one of 2·24 Pa. This should be compared with the value for activated sludge at the same solids' concentration of 37·8 Pa (Riley and Forster, 2001).

It is most likely that ATAD will be considered as a single process. However, TAD may be used either as a single process or as a pretreatment in conjunction with a mesophilic anaerobic digester. The latter option can be thought of as pre-pasteurization, a treatment mode which is sometimes discussed as a distinct and separate process (Brade et al., 2000). The rationale for using this latter option is that the thermophilic stage will kill pathogens. This has certainly been shown to be effective. Pagilla et al. (1996) showed that a TAD/mesophilic dual digestion was able to achieve the criteria for Class A solids. The filtration characteristics of this type of sludge are better than either of the purely thermophilic sludges, with a typical CST value of 750 s. However, its pumping characteristics would be considered as being worse with a yield stress at 40 g SS/litre of 3·56 Pa (Forster, 2002).

Lime treatment
The use of lime as a sludge treatment option is well established, albeit not widely used. However, the need to achieve enhanced treatment standards has resulted in its being re-examined. The addition of lime (CaO) to sludge will generate heat due to the reaction between the lime and the water in the sludge. It will also increase the pH which will generate free ammonia from the ammoniacal-nitrogen in the sludge. The combination of the heat, the high pH and the free ammonia will kill bacteria. The aim is to achieve a pH of more than 12 and it is usually recommended that the lime is added at a dose of 30% on a dry solids basis.

De-watering
After digestion, of whatever type, de-watering may need to be considered. Whether this is necessary and the amount of de-watering

Sludge handling and disposal

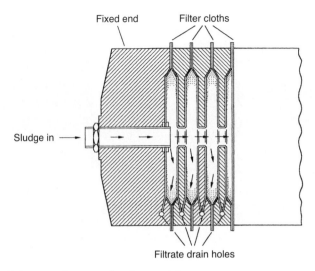

Fig. 8.19. Schematic diagram of a plate and frame press

which is done will depend on the disposal route for the sludge. For example, if it is being injected into agricultural land it may not require any processing other than settlement. If it is going to be incinerated it will require an appreciable amount of de-watering. However, it must be remembered that any de-watering process will generate liquors which will need to be treated and which may constitute an appreciable load, both in terms of BOD and ammoniacal-nitrogen.

Filter press

The plate and frame filter press is one of the most widely used processes for de-watering sludge. It consists of a series of plates which are suspended from a side bar or overhead beam. Each plate is recessed so that chambers are formed between them (Fig. 8.19). Filter cloths are attached to each plate which are then closed and sealed either by screws or hydraulically (Fig. 8.20). Sludge is then pumped in. The pumps also provide the driving force for the filtration process. The pressure used is 600–800 kPa. When the flow of filtrate stops, the plates are separated and the cake is removed. If necessary, the press components are washed with high-pressure water jets and the cycle repeated. Filter pressing is, therefore, a batch process with an expected cycle time of 3–14 hours, depending on the sludge, the cloth and the pressure. The cloths which are used are monofilament fabrics. The choice of filter cloth will depend on how intensely the filter is to be used and, to

Fig. 8.20. Plate and frame press: (a) an individual plate and (b) an assembled unit

some extent, on the sludge/conditioner combination. If only 1–2 runs are envisaged in a day, polypropylene or polyethylene cloths can be used. When a higher number of runs is expected then the more expensive polyamide-11 or polyamide-12 cloths would be the better option. These have a greater durability and resilience than the polyolefin

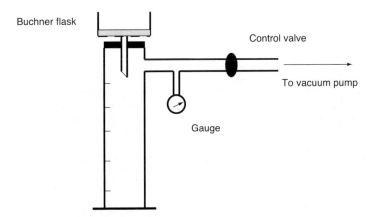

Fig. 8.21. Simple test unit for measuring the specific resistance to filtration

cloths. In addition, they are 'smoother', i.e. the cake is released more readily when the filter is opened at the end of each run.

The rate of filtration is given by

$$\frac{dV}{dt} = \frac{PA^2}{\mu(rVc + RA)} \qquad (8.10)$$

At constant pressure this can be written as

$$\frac{t}{V} = \left(\frac{\mu rc}{2PA^2}\right)V + \frac{\mu R}{PA} \qquad (8.11)$$

where V = volume of filtrate at time t, P = the applied pressure, A = filter area, μ = filtrate viscosity, c = solids' concentration, R = resistance of the filter cloth, r = specific resistance to filtration. The specific resistance to filtration is a measure of how readily a sludge will filter, and the simplest way of measuring in the laboratory is to use the equipment similar to that shown in Fig. 8.21 with a controlled vacuum of 49 kPa. More sophisticated equipment is available but is unlikely to produce a more correct value, and this may not even be necessary. What is required is a measure of the filtrate volume with time so that a plot of (t/V) against V can be obtained. The slope of the line is then used to calculate r. Values of r do vary with the type of sludge. Typically, primary sludge would have a value of 1000×10^{11} to 1500×10^{11} m/kg, whereas the value for activated sludge would be 1000×10^{11} to 4000×10^{11}. The capillary suction time (CST) is an alternative way of assessing the filtration characteristics of sludges (Baskerville and Gale, 1968). The equipment for measuring the CST consists of a rectangle of filter paper (Whatman No. 17; 90×70 mm) placed

Wastewater treatment and technology

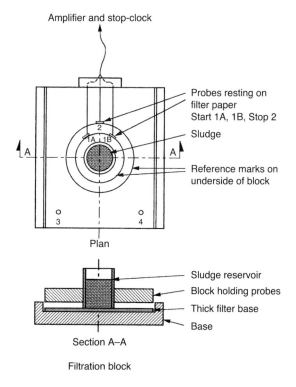

Fig. 8.22. Apparatus for measuring the CST

between two perspex slabs (Fig. 8.22). The upper contains a hole in which the sludge reservoir (10 or 18 mm diameter) is placed. It is also fitted with three stainless steel electrodes located in two concentric circles. These electrodes are connected to an amplifier and a stop-clock. When a

Sludge handling and disposal

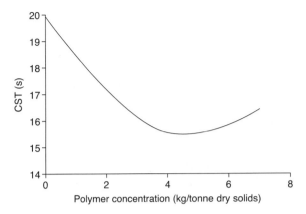

Fig. 8.23. *Typical effect of cationic polymer dosage on the filterability of activated sludge*

sample of sludge is placed in the reservoir, the capillary suction pressure of the paper causes filtration to occur and the filtrate diffuses outwards. When it reaches the two inner electrodes the clock is started, when it reaches the outer electrode the clock is stopped. The CST measurements are a useful way of evaluating the effects of sludge treatments such as conditioning or shear. Values with an 18 mm diameter reservoir can range from 9 s for an iron dosed activated sludge to 70–120 s for activated sludge without any form of treatment.

A specific resistance value of greater than 10×10^{11} to 50×10^{11} m/kg indicates that the sludge has poor filtration characteristics. Most sewage sludges fall into this category and, therefore, require 'conditioning'. This is achieved by dosing the sludge with chemicals such as 'copperas' (ferrous sulphate), ferric chloride or polyaluminium chloride. An alternative, which arguably is the most common choice, is to use cationic organic polymers which have structured molecular weights and charge densities. Whichever conditioner is used, it is essential to use the correct dose. Over-dosing will start to reverse the effect (Fig. 8.23) and will result in unnecessary costs. Under-dosing will result in a less than optimum performance and the longer press times will result in unnecessary running costs.

The membrane filter press is a variant of the conventional plate and frame filter press and uses plates with a flexible diaphragm between the plate and the cloth. This diaphragm can be inflated towards the end of the pressing cycle, using water or air, and, therefore, the cake is compressed more rapidly than would otherwise occur. As such, it can give very much reduced cycle times (Lowe and Shaw, 1992).

Wastewater treatment and technology

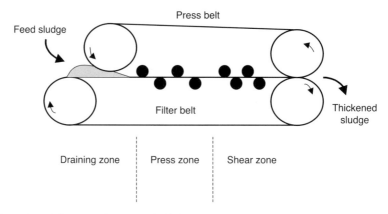

Fig. 8.24. *Schematic diagram of a belt press*

Belt press

With a belt press, sludge is de-watered in a similar way to the gravity belt thickener but using two belts (Fig. 8.24). The lower belt allows the passage of water whereas the upper belt does not. Initially, there

Fig. 8.25. *Sludge being discharged from a belt press*

is a free drainage zone where water is removed by gravity. The two belts then pass over a series of rollers which gradually apply both compression and shear before the de-watered sludge is discharged (Fig. 8.25). The degree of de-watering which occurs will depend on the amount of pressure and shear which is applied but, typically, the sludge produced by a belt press would have a solids' concentration of 15–25%.

Sludge disposal

Disposal to agricultural land

Sewage sludge contains both nitrogen and phosphorus (Table 8.9) and can, therefore, have a beneficial effect when applied to agricultural land (Fig. 8.26). Indeed, the UK Government has stated that 'recovering value from sludge through spreading on agricultural land is the best practicable environmental option for sludge in most circumstances' (Department of the Environment, Transport and the Regions, 2000). In the UK, its use in this way is regulated by the Sludge (Use in Agriculture) Regulations 1989 (amended 1990). There is also a range of Codes of Practice which have been issued by the MAFF. When sewage sludge is applied to agricultural land, as well as there being a need to consider and, where necessary, comply with the requirements of the Safe Sludge Matrix, general stability and the impact of heavy metals must be considered. Stability means that the sludge will not putrify further and give rise to odour complaints. If the sludge has been digested, there should be no problem with this as the putrescible fraction should have been degraded during the digestion process. The conditions of the Safe Sludge Matrix mean that only treated or enhanced treated sludges can be applied to food crops, but it is possible to apply untreated sludges to certain industrial crops such as hemp, which is to be used to produce fibre, and willow and poplar being

Table 8.9. Nitrogen and phosphorus in sewage sludges

Type of sludge	Total N (% dry solids)	NH_3–N (% dry solids)	Phosphorus (% dry solids)
Liquid/digested	1–7	20–70	2–3
Dewatered/digested	1·5–3	<5	–
Limed/primary	–	–	1·1–5·2
Dried	3·5–6	10–15	–

Fig. 8.26. Sludge being injected into farm land

grown for coppicing. However, after 31 December 2005 all use of untreated sludge is to cease.

The dissemination of heavy metals into the environment must also be controlled. There are two categories of heavy metal: those which are phyto-toxic and those which can accumulate in plant tissue and which are toxic when eaten. The phyto-toxic metals are zinc, copper and nickel, and these are frequently considered together as the 'zinc equivalent':

$$\text{Zinc equivalent} = \text{Zn concentration} + 2(\text{Cu concentration}) + 8(\text{Ni concentration})$$

The toxic heavy metals which are of greatest concern are lead, mercury and cadmium. In the UK, the impact of heavy metals is controlled by ensuring that the critical limit in the soil is not exceeded (Table 8.10). This means that the concentration of each proscribed metal, both in the sludge and in the receiving soil, must be known, together with values for the maximum permitted application rate (Table 8.11). This information is then used to calculate the potential application

Table 8.10. Limits for heavy metals in soils (mg/kg dry matter)

	86/278/EEC	France	Germany	Netherlands	UK	
	6 < pH < 7				5 < pH > 7	pH > 7
Cd	1–3	2	1·5	0·8	3	3
Cr	–	150	100	100	–	–
Cu	50–140	100	60	36	80–135[a]	200
Hg	1–1·5	1	1	0·3	1	1
Ni	30–75	50	50	35	50–75[a]	110
Pb	50–300	100	100	85	300	300
Zn	150–300	300	200	140	200–300[a]	450

[a] Depends on pH

rates for each metal

$$\text{Rate} = \left(\frac{\text{soil application limit} - 2\cdot 2 \times (\text{soil concentration})}{\text{sludge concentration}}\right) \times \left(\frac{1000}{\text{time period}}\right) \quad (8.12)$$

The factor of 2·2 converts the metal concentration in the soil expressed as mg/kg to a level expressed as kg/ha, and the time period over which the application will last is taken as 30 years. Thus, if the zinc concentrations of a sludge and the receiving soil were 1650 mg/kg and 40 mg/kg respectively, the potential application rate (using Table 8.11) would be

$$\left(\frac{560 - (2\cdot 2 \times 40)}{1650}\right)\left(\frac{1000}{30}\right) = 9\cdot 5 \text{ kg/ha year} \quad (8.13)$$

The rates for each metal would be derived and the lowest value would determine the actual application rate.

Table 8.11. Maximum permissible limits for addition of heavy metals to soils

Metal	Maximum limit (kg/ha)
Cadmium	5
Copper	280
Lead	1000
Nickel	70
Zinc	560

Wastewater treatment and technology

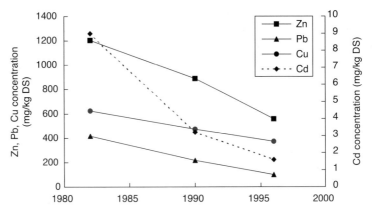

Fig. 8.27. Heavy metal concentrations in sludge

Heavy metals tend to be thought of as resulting from trade effluent discharges. These can be controlled through trade effluent consents and the concentrations of these elements in waste sludge are decreasing (Fig. 8.27). However, appreciable concentrations of copper and zinc come from domestic sources (Hall, 1995) and controlling these inputs at source could be a problem. The European Sewage Sludge Directive (86/278/EEC) does specify maximum limit values in sludge, but as can be seen from Table 8.12 the average heavy metal concentrations in sludges in the member states are well below these limits.

The nitrogen content of the sludge is another constraining factor in the disposal of sludge to agricultural land (Davis, 1999). Leaching of ammoniacal-nitrogen into surface waters needs to be prevented as far as is possible, and the pollution of both surface waters and ground waters must also be minimized. The MAFF Code of Practice on water distinguishes between biosolids which have high available

Table 8.12. Heavy metals in sewage sludges — average concentrations in European Member States

Metal	Maximum concentration (mg/kg)	Range (mg/kg)
Cd	40	0·4–3·8
Cu	1750	39–641
Hg	25	0·3–3
Ni	400	9–90
Pb	1200	13–221
Zn	4000	142–2000

Sludge handling and disposal

nitrogen, such as liquid digested sludge, and those which have low available nitrogen. Sludge cake, lime treated sludge and thermally dried sludge fall into this latter category. The Code of Practice recommends that sludge applications should supply not more than 250 kg total nitrogen per hectare per year or not more than 500 kg total nitrogen per hectare every other year for solids with a low nitrogen availability. This must be considered in conjunction with the heavy metal limitations. The Nitrate Directive (91/676/EEC) also has implications for the use of sludge on land. It requires that nitrate-sensitive zones must be identified and means that the application of sludges in these areas will be restricted.

The phosphorus content of sludge also needs to be considered. The Water Code of Practice recommends that the total phosphorus input should not exceed the amount removed by the crops which are being grown so that the leaching of phosphorus into surface waters, and the consequent risk of eutrophication, is reduced. The phosphorus input from sludge being applied to meet the 250 kg total nitrogen per hectare, per year limit would range from about 100 to 350 kg P_2O_5 per hectare depending on the type of sludge being used (Hickman, 1999). A wheat crop of 8 tonne/hectare would only use 62 kg/hectare. Therefore, if phosphorus accumulation is to be avoided, sludge application has to be restricted.

Incineration

Incineration is an option for the ultimate disposal of contaminated sludges for which there is no other sink. In some cases, however, it may be the most cost-effective option, regardless of the state of contamination. The sludges should be as dry as possible as there is no point in using fuel to evaporate water. This means using a filter press or, possibly, a centrifuge. In some cases, the cakes from either the press or a centrifuge may be dried further using waste heat from the incinerator itself.

There are a number of incinerator designs which may be considered, but the trend in 2003 is to use fluidized bed incinerators (Fig. 8.28). This is the result of modern de-watering techniques which can produce a cake with consistent quality compatible with autothermal operation of the incinerator. Autothermicity relates to the organic content of the sludge. The higher it is, the higher will be the calorific value of the sludge and, if it is sufficiently high, the system will be autothermal and not require any additional fuel. The calorific values of some typical sludges are given in Table 8.13. As a guideline, a

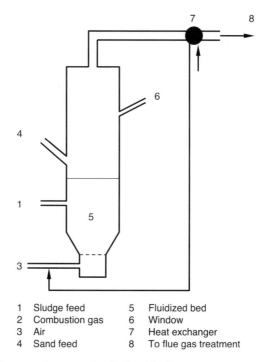

Fig. 8.28. Schematic diagram of a fluidized bed incinerator

1	Sludge feed	5	Fluidized bed
2	Combustion gas	6	Window
3	Air	7	Heat exchanger
4	Sand feed	8	To flue gas treatment

sludge de-watered to a dry solids' concentration of about 33% would generate autothermal conditions if the combustion air were preheated. If the air was not pre-heated, a solids' concentration of about 45% would be needed for autothermicity. The inorganic residues are discharged with the flue gases as fly ash and are removed in the gas cleaning phase. In the current climate of growing public concern about the incineration of any waste, gas cleaning is of major significance and sophisticated techniques do have to be used. The degree of sophistication is such that the emission control facilities can account for about half the mechanical and electrical capital costs of a waste to energy

Table 8.13. Calorific values of thickened sewage sludges

Type	Dewatered	CV (kJ/kg ash-free DS)
Primary/humus	Belt press	13 600–16 700
Primary activated	Centrifuge/drier	16 600–18 900
Primary/humus/activated	Belt press/drier	16 100–18 900

Table 8.14. *Environment Agency stack emissions limits for the Shell Green incineration plant (Belshaw, 2000)*

Parameter	Daily mean	½ hour mean	Spot sample
Particulates (mg/m^3)	10	30	
HCl (mg/m^3)	10	60	
Sulphur oxides (mg/m^3)	50	200	
Nitrogen oxides (mg/m^3)	200	400	
Carbon monoxide (mg/m^3)	50	100	
Organic carbon (mg/m^3)	10	20	
HF (mg/m^3)	1	4	
Mercury (mg/m^3)			0·05[a]
Cadmium (mg/m^3)			0·05[a]
Total metals[b] (mg/m^3)			0·5[a]
Dioxins and furans (μg^3/m)			0·1[c]
Odour (units/s)			1220[c]
Odour (units/m^3)			100[c]

[a] Quarterly sample
[b] Arsenic, copper, chromium, lead, manganese, nickel, tin, cobalt
[c] Bi-annual sample

incineration plant. The stack emissions from a sludge incinerator are subject to authorization by the Environment Agency (Table 8.14) and this could include an odour standard if the site was in a sensitive location (Belshaw, 2000). Electrostatic precipitation is one technique for removing solids and is the option used at fluidized bed incinerators operated by Yorkshire Water where a removal rate of some 95% is achieved. The process entails using electrostatic forces of attraction to capture wet and dry particles. The main operating advantage is the low-pressure drop across the system. It is a process which will work at temperatures up to c. 500 °C. Typical throughputs are more than 1000 m^3/s and the operating voltage is 30 000–100 000 V. In this field, ions are generated from the negative electrode and become attached to the particles. These are then attracted to the positive electrode where they become trapped.

The efficiency (E), which tends to be high, is given by

$$E = 1 - \left[\exp\left(\frac{-Aad_p}{Q}\right)\right] \tag{8.14}$$

where A = collection plate area (m^2), a = constant, d_p = particle diameter (m), Q = gas flow rate (m^3/s). The ash is approximately 55% silica but will also contain the heavy metals present in the

Wastewater treatment and technology

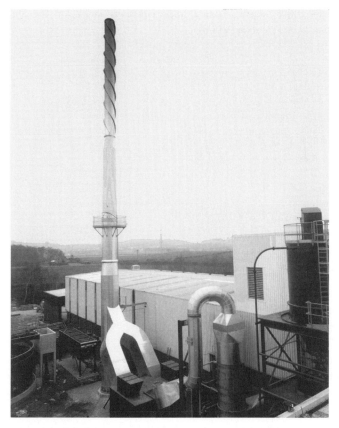

Fig. 8.29. The incinerator at Roundhill (courtesy of Severn Trent Water)

sludge. After removal, it is collected and stored before disposal, usually to a land fill site. Because of the presence of the heavy metals it is possible that the ash could be classified as hazardous waste. If this happens the number of sites available for disposal becomes limited.

Severn Trent Water use fluidized bed incineration at their Roundhill Sewage Treatment Works (Fig. 8.29). It has an operating temperature of more than 850 °C with an average retention time of about 4 s. The sludge comes from two sources: filter cake at about 25% DS, or digested sludge at about 3% DS. The liquid sludge is de-watered using polyelectrolyte treatment and centrifugation. This is mixed with the filter cake and pre-dried, using pressurized hot water (180 °C), to a solids' concentration of 32%. Screenings are also incinerated. The air from the solids handling area, which can be odorous, is extracted and used in the incinerator. This reduces the risk of odour nuisance. The exit

Sludge handling and disposal

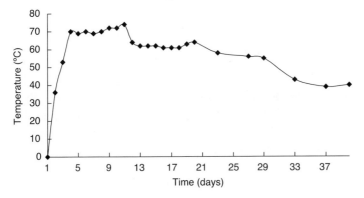

Fig. 8.30. Temperature profile during composting

gases are used to pre-heat the incoming air to 650 °C and then the water for the pre-drier. The gas cleaning stream is made up of

- spray drier/adsorber;
- fabric filters;
- wet scrubber.

Landfill

The EC Directive on hazardous wastes does not classify sludges from municipal treatment works as hazardous unless they contain listed toxic substances or contain heavy metals in concentrations which makes them unsuitable for agricultural use. Non-hazardous sewage sludges can, therefore, be disposed of in landfill sites. However, landfill is not widely used in the UK as a disposal option (see Table 8.1). The requirements of the EC Directive on landfill are that the minimum solids' concentration of the sludge should be 35%. This, essentially, requires the sludge to have been de-watered by filter pressing. Some countries in the EU, for example the Netherlands, also have a restriction on the organic content of material which can be accepted in landfill sites.

Composting

Composting involves the biological degradation and stabilization of organic matter under aerobic conditions. As with the ATAD process (see earlier) the heat generated by the biological processes should enable thermophilic temperatures to be reached (Fig. 8.30). For the composting to proceed rapidly, the solids should have a carbon:nitrogen ratio in the range 25:1 to 30:1. Sewage sludge will have a C:N ratio of about 10:1 and, therefore, needs to be mixed with some other material which has a higher ratio. Straw and woody-type wastes are typical of such a material having C:N ratios of about 80–100:1 and 500:1 respectively (Biddlestone et al., 1987). Straw can be difficult to mix uniformly with the sludge whereas the use of woodchips and shredded green waste will achieve a more intimate mixture (Bowman and Durham, 2002). The composting process also requires a moisture content of more than 40%, the optimum being 60%. At higher moisture contents the free pore space is reduced to such an extent that anaerobic conditions can prevail. The actual process can be done in turned piles or windrows, aerated static piles or in composting reactors (Carroll et al., 1993). Windrows (1–2 m high and 4–6 m wide) are generally constructed on concrete bases and turned mechanically to enable all the mixture to spend some time in the centre of the pile where the high temperatures prevail. In the work reported by Bowman and Durham (2002), the windrows consisting of sludge and straw in a ratio of 8:1 (by weight) were turned three times a week for five weeks. This was followed by a maturation period of several months. Static piles, which take up less area, are, as the name suggests, not turned but are subjected to forced aeration by either sucking or blowing air into the pile using channels or pipes laid beneath the piles. The work by Bowman and Durham (2002) suggests that piles with blown air operate at lower temperatures than those in which the air is sucked into the pile. Because static bays occupy less area than turned piles, they can be located in enclosed units with the result that odour control (see Chapter 5) is more easily accomplished. The study by Thames Water, which assessed the main types of in-vessel composting systems, reached the conclusion that the agitated bay system was the preferred option for large sludge treatment centres (Bowman and Durham, 2002). These are located in odour-controlled buildings and consist of long bays which contain the compost mix to a depth of 2–3 m. The mix is turned and moved along the bay mechanically. This provides aeration, mixing and a continuous flow of material.

Traditionally, composting of sewage sludge has not been widely practised in the UK. However, the current concepts of sustainability and recycling may well act as drivers for change. Furthermore, the temperatures generated during composting will kill pathogens. Certainly, in one trial using an aerated static pile composting, the temperature was maintained at 55–60 °C and the faecal indicator organisms were reduced from 10^7/g to 10^2/g (Pereira-Neto et al., 1986).

Thermal drying

Thermal drying of sludge may be considered either as a treatment in itself or as a precursor to some further treatment. Any type of sludge may be thermally dried but would normally be a de-watered one. Some consideration might be given to digesting it, prior to drying, so that the mass of solids was reduced, and to provide fuel for the drying process in the form of biogas. There are a range of different types of drying process (Lowe, 1995), the main differences being in the way in which the heat is transferred to the sludge. The options available are direct drying using hot air or indirect drying which involves the use of heat exchangers and a heating medium such as oil or steam. The types of dryer which are available include flash dryers, revolving drum dryers and fluidized beds. Wherever possible, heat recovery systems are used to increase energy efficiency. The end product is a stable, granular, dust free solid (Fig. 8.31) which does not contain any pathogens or parasites. Typically, it would contain 90–95% dry solids (Table 8.15) and would occupy about 7% of the original sludge volume. It would have a sufficiently rigid structure to withstand storage and transportation.

Although thermal drying has been carried out in the USA since the 1980s, in 2003 there were only two dryers in the UK. The larger of these is that operated by Wessex Water at the Avonmouth Sewage Treatment Works in Bristol. It is a Combi process capable of

Fig. 8.31. Thermally dried sludge granules (courtesy of Wessex Water)

Table 8.15. Typical composition of Biogran Natural

Component	Approximate composition (%)
Organic matter	50
Total nitrogen	3·3
Total phosphate	4·4
Water soluble phosphate	0·5
Water soluble nutrients	0·4
Iron and trace elements	1
Total potash (as K_2O)	0·2

handling 200 000 tonnes of wet anaerobically digested sludge (about 4% dry solids) per year (Fig. 8.32). After the preliminary de-watering stage, the sludge is mixed with some of the dried granular sludge to give a sludge containing about 50% dry solids. This is fed to a rotary drum dryer where it is contacted with air heated to about 450 °C. The dried sludge then passes to a sizer which grades the granules prior to bagging. The selected size range is 2–4 mm and under- or over-sized granules are recycled. The air from the drum dryer is filtered and recycled. The dryer is operated under a slight vacuum, thus reducing the risk of odours or other gas emissions. The end product, Biogran Natural, is being actively marketed as a slow-release soil conditioner. As such, Wessex Water are viewing their sludge drying process as a technique for producing a useful material rather than one which makes the sludge more convenient for disposal.

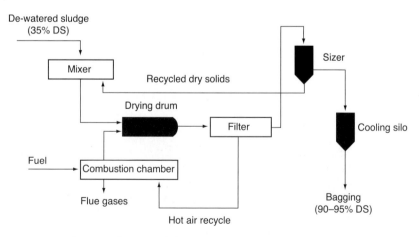

Fig. 8.32. Schematic diagram of the Combi sludge drying process

References

Aitken, M. D. and Mullennix, R. W. (1992) Another look at thermophilic anaerobic digestion of wastewater sludge, *Wat. Environ. Res.*, **64**, 915–19.

Alatiqi, I. M., Hamoda, M. F. and Dadkhah, A. A. (1998) Kinetic analysis of anaerobic digestion of wastewater sludge, *Wat. Air Soil Pollut.*, **107**, 393–407.

Andrew, P. R. and Salt, A. (1987) The Bury sludge digestion plant: early operating experiences, *J. IWEM*, **1**, 22–6.

Anon. (1999) Safe Sludge Matrix: Guidelines for the Application of Sewage Sludge to Agricultural Land, ADAS (AMPU 1234/C).

Baskerville, R. C. and Gale, R. S. (1968) A simple automatic instrument for determining the filterability of sewage sludges, *Wat. Pollut. Control*, **67**, 233–41.

Belshaw, C. (2000) The Mersey Valley sludge story: From sea to Shell Green, *J. CIWEM*, **14**, 193–9.

Biddlestone, A. J., Gray, K. R. and Day, C. A. (1987) Composting and straw decomposition, in *Environmental Biotechnology*, Forster, C. F. and Wase, D. A. J. (eds), Ellis Horwood, Chichester, pp 135–75.

Bowman, L. and Durham, E. (2002) A biosolids composting challenge: Meeting demand for a peat-free horticultural-grade product, *J. CIWEM*, **16**, 105–10.

Brade, C. E. and Noone, G. P. (1981) Anaerobic sludge digestion — Need it be expensive? Making more of existing resources, *Wat. Pollut. Control*, **80**, 70–94.

Brade, C. E., Noone, G. P., Powell, E., Rundle, H. and Whyley, J. (1982) The application of developments in anaerobic digestion within Severn Trent Water Authority, *Wat. Pollut. Control*, **81**, 200–19.

Brade, C. E., Harrison, D., Cumiskey, A. and Dawson, M. (2000) Pre-pasteurization and anaerobic digestion — A low cost route to Class A biosolids, *73rd Annual Conf. Exposit. Water Quality and Wastewater Treatment*, Wat. Environment. Fed., California, USA, Oct.

Bruce, A. M. and Evans, T. D. (2002) *Sewage Sludge Disposal: Operational and Environmental Issues. A Review of Current Knowledge*, FR/R0001, Foundation for Water Research.

Bruce, A. M., Pike, E. B. and Fisher, W. J. (1990) A review of treatment process options to meet the EC sludge Directive, *J. CIWEM*, **4**, 1–13.

Bryant C. (1993) Sludge standards set numerical limits, *Pollut. Engng.*, **25**, 48–50.

Carroll, B. A., Caunt, P. and Cunliffe, G. (1993) Composting sewage sludge: Basic principles and opportunities in the UK, *J. IWEM*, **7**, 175–81.

Christensen, J. R., Sorenson, P. B., Christensen, G. L. and Hansen, J. A. (1993) Mechanisms for overdosing in sludge conditioning, *J. Env. Eng.*, **119**, 159–71.

Clark, P. B. and Nujjoo, I. (2000) Ultrasonic sludge pretreatment for enhanced sludge digestion, *J. CIWEM*, **14**, 66–71.

Cohen, A., Breure, A. M., van Andel, J. C. and van Deursen, A. (1982) Influence of phase separation on the anaerobic digestion of glucose. II — Stability and kinetic responses to shock loading, *Water Res.*, **16**, 449–54.

Cohen, D. B. (1977) A comparison of pure oxygen and diffused air digestion of waste activated sludge, *Prog. Water Technol.*, **9**, 691–702.

Davarinejad, A. and Voutchkov, N. (1994) Effect of sludge blend on dewatering characteristics, *Proc. National Conf. Environ. Engng.*, ASCE, 273–80.

Davis, R. B. (1996) The impact of EU and UK environmental pressures on the future of sludge treatment and disposal, *J. CIWEM*, **10**, 65–9.

Davis, R. B. (1999) The impact of EU and UK environmental pressures on the future of sludge treatment and disposal, *Sludge 9 — Seminar on Developments in Sludge Treatment and Disposal*, University of Surrey, Sept.

Department of the Environment (1996) *Code of Practice for Agricultural Use of Sewage Sludge*, HMSO, London, 1989, Revised Edition.

Department of the Environment, Transport and the Regions (2000) *Waste Strategy 2000 for England and Wales*, DETR.

Dick, R. I. and Ewing, B. B. (1967) The rheology of activated sludge, *J. Wat. Pollut. Control Fed.*, **39**, 543–60.

Eastman, J. A. and Ferguson, J. F. (1981) Solubilisation of particulate organic carbon during the acid phase of anaerobic digestion, *J. Wat. Pollut. Control Fed.*, **53**, 352–66.

Farrell, P. D., Nair, A., Palmer, S. and Davies, W. J. (2001) Control and stabilisation of odorous sludge from Fleetwood sewage treatment works, *J. CIWEM*, **15**, 46–50.

Forster, C. F. (2002) The physico-chemical characteristics of sewage sludge, *Enz. Microb. Technol.*, **30**, 340–5.

Frost, R. C. (1983) How to design sewage sludge pumping stations. *WRc Technical Report TR185*, Water Research Centre, Medmenham.

Haarhoff, J. and Bezuidenhout, E. (1999) Full-scale evaluation of activated sludge thickening by dissolved air flotation, *Water SA*, **25**, 153–66.

Hall, J. E. (1995) Sewage sludge production, treatment and disposal in the European Union, *J. CIWEM*, **9**, 335–43.

Hanaki, K., Matsuo, T., Nagase, M. and Tabata, Y. (1987) Evaluation of effectiveness of two-phase anaerobic digestion process degrading complex substrate, *Wat. Sci. Technol.*, **19**, 311–22.

Hickman, G. A. W. (1999) Sustainable beneficial biosolids recycling to land — Future issues, *Sludge 9 — Seminar on Developments in Sludge Treatment and Disposal*, University of Surrey, Sept.

Holt, S. and Hudson, J. A. (1999) Submerged combustion for prepasteurisation of sewage sludges, *Sludge 9 — Seminar on Developments in Sludge Treatment and Disposal*, University of Surrey, Sept.

Hoyland, G., Dee, A. and Day, M. (1989) Optimum design of consolidation tanks, *J. IWEM*, **3**, 505–16.

Hudson, J. A. (1995) Treatment and disposal of sewage sludge in the mid-1990s, *J .CIWEM*, **Centenary Issue**, 93–100.

Hudson, J. A. and Lowe, P. (1996) Current technologies for sludge treatment and disposal, *J. CIWEM*, **10**, 436–41.

Jewell, W. J. and Kabrick, R. M. (1980) Autoheated aerobic thermophilic digestion with aeration, *J. Wat. Pollut. Control Fed.*, **52**, 512–23.

Kidson, R. J. and Ray, D. L. (1984) Pasteurisation by submerged combustion together with anaerobic digestion, in *Sewage Sludge Stabilisation and Disinfection*, Bruce, A. M. (ed), Ellis Horwood, Chichester, pp 399–411.

Kiyohara, Y., Miyahara, T., Mizuno, O., Noike, T. and Ono, K. (2000) A comparative study of thermophilic and mesophilic sludge digestion, *J. CIWEM*, **14**, 150–4.

Komatsu, T., Hanaki, K. and Matsuo, T. (1991) Prevention of lipid inhibition in anaerobic processes by introducing a 2-phase system, *Wat. Sci. Technol.*, **23**, 1189–200.

Lewis-Jones, R. and Winkler, M. (1991) *Sludge Parasites and Other Pathogens*, Ellis Horwood, Chichester.

Li, Y. Y., Noike, T., Katsumata, K. and Koubayashi, H. (1996) Performance analysis of the full-scale egg-shaped digester treating sewage sludge of high concentration, *Wat. Sci. Technol.*, **34(3–4)**, 483–91.

Lowe, P. (1995) Developments in the thermal drying of sewage sludge, *J. CIWEM*, **9**, 306–13.

Lowe, P. and Shaw, D. (1992) Development of the membrane filter press for the processing of sewage sludge, *Wat. Sci. Technol.*, **25(4–5)**, 297–305.

Mayhew, M., Le, S., Ratcliffe, R., Brade, C. and Harrison, D. (2001) Advances in pathogen destruction with enzymic hydrolysis, *6th European Biosolids and Organic Residues Conf.*, Aqua-Enviro, Wakefield, UK, Nov.

Moeller, G. and Torres, L. G. (1997) Rheological characterisation of primary and secondary sludges treated by both aerobic and anaerobic digestion, *Bioresource Technol.*, **61**, 207–11.

Mosey, F. E. (1976) Assessment of the maximum concentration of heavy metals in crude sewage which will not inhibit the anaerobic digestion of sludge, *Wat. Pollut. Control*, **75**, 10–20.

Nielsen, B. and Petersen, G. (2000) Thermophilic anaerobic digestion and pasteurisation. Practical experience from Danish wastewater treatment plants, *Wat. Sci. Technol.*, **42(9)**, 65–72.

Oles, J., Dichtl, N. and Niehoff, H-H. (1997) Full-scale experience of two stage thermophilic/mesophilic sludge digestion, *Wat. Sci. Technol.*, **36**, 449–56.

Pagilla, K. R., Craney, K. C. and Kido, W. H. (1996) Aerobic thermophilic pretreatment of mixed sludge for pathogen reduction and Nocardia control, *Wat. Environ Res.*, **68**, 1093–8.

Papacrivopoulos, D. (1985) *Pumped Desludging of Primary Sedimentation Tanks at Minworth*, MSc Thesis, University of Birmingham.

Pereira-Neto, J. T., Stentiford, E. I. and Smith, D. V. (1986) Survival of faecal indicator micro-organisms using the aerated static pile system, *Waste Mangt & Res.*, **14**, 397–406.

Price, G. J. and Alder, M. (1985) The economic advantages of sludge thickening at Avonmouth sewage treatment works, *Wat. Pollut. Control*, **84**, 394–406.

Quarmby, J., Scott, J. R., Mason, A. K., Davies, G. and Parsons, S. A. (1999) The application of ultrasound as a pre-treatment for anaerobic digestion, *Environ. Technol.*, **20**, 1155–62.

Riley, D. and Forster, C. F. (2001) The physico-chemical characteristics of thermophilic aerobic sludges, *J. Chem. Tech. Biotech.*, **76**, 862–6.

Roberts, R., Le, S. and Forster, C. F. (1999) A thermophilic/mesophilic dual digestion system for treating waste activated sludge, *J. Chem. Tech. Biotech.*, **74**, 445–50.

Speece, R. E. (1983) Anaerobic biotechnology for industrial wastewater treatment, *Env. Sci. Technol.*, **17**, 416A–27A.

Strauch, D. (1998) Pathogenic micro-organisms in sludge: anaerobic digestion and disinfecting methods to make sludge usable as a fertiliser, *Eur. Water Mngt.*, **1**, 102–11.

Swinwood, J. F. and Wilson, B. K. (1990) Sewage-sludge pasteurization by gamma radiation: a Canadian demonstration project — 1988–91, *Radiation Physics and Chemistry*, **35**, 461–4.

Tanaka, S. and Matsuo, T. (1986) Treatment characteristics of the two-phase anaerobic digestion system using an upflow filter, *Wat. Sci. Technol.*, **18**, 217–24.

Thomas, V. K., Adraktas, D. V. and Taylor, C. (2000) Provision of sewage and sludge treatment for Athens, *Wat. Sci. Technol.*, **41(9)**, 37–43.

Tolan, D. J., Bennett, D. and Kwiecinski, J. V. (1989) Blackburn Meadows sewage treatment works: Sludge incineration into the 1990s, *J. IWEM*, **3**, 239–49.

van Lier, J. B., Tilche, A., Ahring, B. K., Macarie, H., Moletta, R., Dohanyos, M., Hulshoff Pol, L. W., Lens, P. and Verstraete, W. (2001) New perspectives in anaerobic digestion, *Wat. Sci. Technol.*, **43(10)**, 1–18.

Walker, S. (1996) Energy from waste in the sewage treatment process, *IEE Conf. Proceedings*, 73–5.

WEF (1992) *Manual of Practice No. 8, Design of Municipal Wastewater Treatment Plants, Vol II*, USA Books Inc.

Westoll, T. G. (1999) Driffield sewerage and sewage-treatment projects, *J. IWEM*, **13**, 121–6.

Williams, R., Padley, P. J. and Knight, A. J. (1993) Aspects of submerged combustion as a heat exchange method, *Chem. Eng. R and D*, **71**, 308–9.

Zhao, Q. L. and Kugel, G. (1996) Thermophilic/mesophilic digestion of sewage sludge and organic wastes, *J. Environ. Sci. Health*, **31**, 2211–31.

Zickefoose, C. and Hayes, R. B. (1976) *Anaerobic Sludge Digestion: Operations Manual*, EPA 430/9-76-001.

9
Anaerobic digestion of wastewaters

Introduction
The anaerobic digestion of liquid wastewaters focuses on degrading COD rather than volatile solids, as is the case in the treatment of sludge (Chapter 8). The components of the two types of waste which are degradable are similar: proteins, carbohydrates and lipids. Therefore, the biochemical pathways are, essentially, the same. However, it could be argued that a better understanding of the microbial ecology is needed when dealing with the treatment of liquid wastes.

Microbiology
There are probably two aspects which need to be considered: the growth rates of all the species and the nature of the methanogens. The growth rates are important if modelling of the processes is to be attempted. Table 9.1 shows some typical values of the kinetic constants.

The methanogens are not procaryotes (bacteria) but belong to the *Archaebacteria*. As such, they have different cell walls and lipids. They also contain specific co-enzymes and co-factors, some of which can fluoresce under ultraviolet light, a factor which can be used to examine them microscopically. The existence of these co-enzymes and co-factors means that the methanogens have a requirement for specific heavy metals. The most important are molybdenum (0·1 mg/litre), iron (0·12 mg/litre), cobalt (0·6 mg/litre) and nickel (6·0 mg/litre). About seven acetoclastic species have been identified in anaerobic sludges, the most common being *Methanosarcina* and *Methanosaeta*. The latter used to be known as *Methanothrix* (Patel and Sprott, 1990). The

Table 9.1. Kinetic constants for the biomass in anaerobic digestion

Type	μ_{max} (day^{-1})	K_S (kg COD/m^3)	Y_{max} (kg VSS/kg COD)
Acidogenic bacteria	2·0	0·2	0·15
Methanogenic bacteria	0·4	0·05	0·03
Methanosarcina[a]	0·3	0·18–0·60	–
Methanosaeta[a]	0·1	3·0–4·8 × 10^{-3}	–

[a] Relates to acetate growth

morphology of these species is quite characteristic. *Methanosarcina* are irregular clusters of semi-spherical cells and *Methanosaeta* are filamentous (Figs 9.1 and 9.2). The other methanogens are rods or cocci. *Methanosarcina* can utilize hydrogen, formate, methanol and methylamine as well as acetate, whereas *Methanosaeta* can only utilize hydrogen and acetate. The growth characteristics of these two species (Table 9.1) mean that *Methanosaeta* will out-compete *Methanosarcina* when acetate concentrations are low. Other species found in the anaerobic biomass include *Methanobrevibacter* spp., *Methanospirillum* spp. and *Syntrophobacter* spp.

As well as competition between the two acetoclastic species there may be competition between the methanogens in general and the sulphate reducing bacteria (SRBs). This will occur when there are appreciable concentrations of sulphate in the wastewater. Both the SRBs and the methanogens compete for the same electron donors, hydrogen and acetate, and, at low acetate concentrations, the SRBs

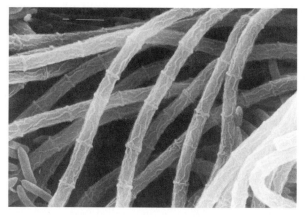

Fig. 9.1. Scanning electron micrograph of Methanosaeta (bar = 1 μm)

Fig. 9.2. Scanning electron micrograph of a biomass dominated by Methanosarcina

will out-compete the methanogens. This means that not only will there be a reduction in the quantity of methane produced, but there also will be contamination of the biogas with hydrogen sulphide. This can result in odour problems and cause corrosion if the biogas is being used as a fuel. The $COD:SO_4$ ratio is the critical aspect and a ratio of less than 1·7–2·7 will favour the SRBs (Choi and Rim, 1991).

Reactor design

Early attempts to use anaerobic digestion for the treatment of industrial wastewaters resulted in hydraulic retention times which were not compatible with the needs of industry. Effectively, they were too long. To overcome this, reactor designs were developed which separated the hydraulic retention time from the solids' retention time. This had the effect of retaining a high concentration of biomass in the reactor which, in turn, enabled the COD in the wastewater to be degraded more rapidly.

Contact digester

The immobilization of the biomass in a contact digester is achieved by having a phase separation/biomass recycle system similar to that in the activated sludge process. This can be a settlement tank or a membrane (Chapter 5). If a settlement tank is used, there needs to be some process

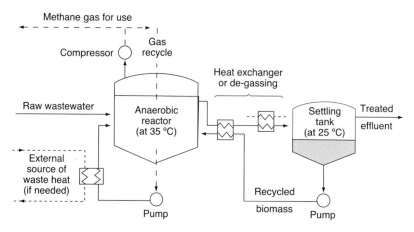

Fig. 9.3. *Schematic diagram of a contact digester*

for stopping gas production which would otherwise cause sludge particles to be lifted to the surface, in much the same way as they are when de-nitrification occurs in the final tank of an activated sludge plant. Techniques which have been used include cooling to about 20 °C or de-gassing (Fig. 9.3). The biomass densities which can be achieved depend on the strength of the waste being treated. With high-strength (more than 20 g COD/litre) feeds, solids' concentrations of 20–30 g VSS/litre have been reported (Barnes and Fitzgerald, 1987).

The contact digester has no internal fittings and, therefore, is very suitable for dealing with wastewaters which contain high concentrations of suspended solids. It is also a system which is well mixed and this is one reason for the removal efficiencies which can be achieved (Table 9.2). Loading rates tend to be in the range of 1–10 kg COD/m^3 per day with retention times of 2–5 days and, provided that the settlement phase can be managed, the process is generally robust.

Table 9.2. *Performance data for contact digesters*

Waste	Volume (m^3)	OLR (kg/m^3 per day)	COD removal (%)
Starch	900	3·6	65
Yeast	1900	2·8–3·9	77–82
Citric acid	10 000	1·3–4·0	75–83
Pectin	3000	4·2	80

Table 9.3. *Characteristics of a contact digester treating brewery wastewater* (Wheatley et al., 1997)

Parameter	
Size	2000 m^3
Dimensions	14 m high
Clarifier volume	900 m^3
Flow	1200 m^3 per day
OLR	2·5 kg/m^3 per day
HRT	36 h
Influent COD	4000 mg/litre
Influent SS	600 mg/litre
COD Removal	80–90%
Biomass concentration	1·0%
Recycle ratio	1 : 1

There have been a number of surveys on this type of reactor (Shore *et al.*, 1984; Smith *et al.*, 1988) and one of the more detailed was that carried out on the digester treating the wastewater from the Wrexham Lager plant (Wheatley *et al.*, 1997). The digester (Table 9.3) was built in 1986 at a capital cost of £375/m^3 of the volume of the digester and a survey over six years showed that, in general, the COD removal was better than 90%.

Anaerobic filter

The anaerobic filter can be operated in either of two modes: upflow or downflow. Probably the upflow system is the more common. Both systems use plastic media which is fully submerged and, typically, the packing would occupy 60–70% of the reactor volume (Fig. 9.4). The downflow option would be more likely to use an orientated or tubular packing, whereas an upflow filter would use orientated or random packing. Because of the counter-current flow of the gas and liquid in the downflow system, mixing would be more intense than in an upflow filter. The type of biomass and the way it is immobilized will also be different in the two designs. In the upflow filter it exists partly as a biofilm on the media and partly in a flocculated form trapped in the voids. The downflow filter is a thin film process and, as with most biofilm processes, surface roughness of the support media will enhance the attachment of the film and, therefore, the start up. Also, the surface to volume ratio of the support matrix is important in determining the amount of biomass. This, in turn, affects the loading

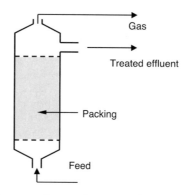

Fig. 9.4. Schematic diagram of an upflow anaerobic filter

rate which can be applied. Tables 9.4 and 9.5 give some typical performance data for downflow and upflow filters respectively and shows that they are operated at higher loading rates than contact digesters.

Upflow sludge blanket reactor

In this process, the wastewater being treated is passed upwards through a bed of sludge particles where the main biochemical reactions occur (Fig. 9.5). If good biomass retention is to be achieved, the sludge has to have good settling particles and be sufficiently aggregated to

Table 9.4. Performance of downflow filters (Barnes and Fitzgerald, 1987)

Wastewater	Loading rate (kg COD/m^3 per day)	COD removal (%)
Bean processing	9–18	87
Piggery	6	70
Cheese production	8	60–70
Volatile fatty acids	5–18	80
Rum distillation	8–10	65–70

Table 9.5. Performance of upflow filters

Wastewater	Loading rate (kg COD/m^3 per day)	COD removal (%)
Wheat starch	9·5	66
Jam processing	10·5	90
Dairy wastes	10·4	80
Food canning	8–16	58–64

Anaerobic digestion of wastewaters

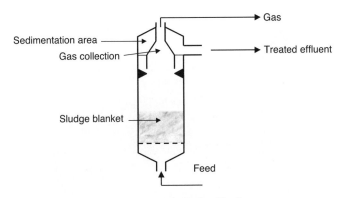

Fig. 9.5. *Schematic diagram of an upflow sludge blanket reactor*

overcome the hydraulic shear of the upward flowing liquid. Upflow anaerobic sludge blanket (UASB) systems, therefore, rely on the formation of a granular sludge. The granules which are formed in UASB reactors can have diameters of 0·14–5 mm (Fig. 9.6) and appear to be formed by bacterial selection. When examined with an electron microscope, holes can often be seen in the granules (Fig. 9.7). These are thought to be channels for gas release (Morgan et al., 1991). Substrates which are degradable only slowly can be deposited on the surface of the granules. An example of this is the accumulation of lipids during the treatment of ice-cream wastewater (Morgan et al., 1990) (Fig. 9.8). This caused the specific gas yield to be reduced significantly. The importance of granules in the operation of the UASB process has led to an appreciable amount of work being

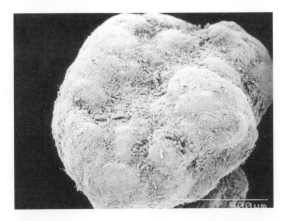

Fig. 9.6. *Scanning electron micrograph of a granule from an upflow sludge blanket reactor treating sugar refinery wastewater*

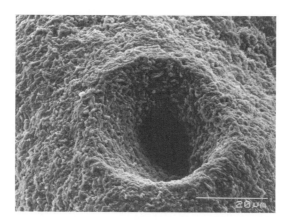

Fig. 9.7. Scanning electron micrograph of a granule showing gas cratering (Morgan et al., 1991)

done to elucidate both their structure and the way in which they are formed. Details of this work are given in the review by Schmidt and Ahring (1996). The filamentous methanogen, *Methanosaeta*, formerly known as *Methanothrix* (Patel and Sprott, 1990), appears to play a significant role, as does extracellular polymeric material (Schmidt and Ahring, 1996). Granules have a high settling velocity (0·012 m/s) and certainly have the ability to withstand relatively high hydraulic shear. Granular sludge is also reported as having higher specific activities than non-granular sludge; 4 kg COD removed/kg VSS compared with 1·8 kg/kg (Hulshoff-Pol et al., 1982).

The other main feature of these reactors is the solid/gas separation device which is located at the top of the reactor. Table 9.6 gives some

Fig. 9.8. Scanning electron micrograph of a granule showing the accumulation of fatty material on the granule surface (Morgan et al., 1990)

Table 9.6. Performance data for UASB reactors

Waste	Flow (m³/d)	OLR (kg/m³ per day)	HRT (h)	COD removal (%)	Gas yield (m³/d)
Paper	15 000	9·6	4	68	6500
Paper	6240	12·0	7·5	87	4800
Brewery	6100	8·6	8·4	80	5800
Brewery	1350	7·0	5·3	77	360
Potato	1000	13·5	11·4	81	2000
Coffee	650	5·2	16·0	70	660

typical performance data for UASB reactors and shows that they are operated at higher loading rates than contact digesters and anaerobic filters.

The hybrid reactor is an extension of the upflow sludge blanket reactor which has packing media above the sludge blanket (Fig. 9.9). Thus, it combines the attributes of both the UASB and the anaerobic filter. However, there is no need for the sludge to be in a granular form. It also means that there is a greater mass of biomass retained within the reactor.

The expanded granular sludge bed (EGSB) is another extension of UASB technology (Fig. 9.10). Effectively, it is a fluidized bed reactor which uses a granular sludge rather than a biofilm on a support material and, as such, can operate at high loading rates; typically, 15–30 kg COD/m³ per day. One of the most important features of the Biobed® EGSB is that the three-phase settler at the top of the reactor will tolerate higher upflow velocities (up to 15 m/h) than those used in the UASB (1·0–1·5 m/h). There is also a difference in the heights

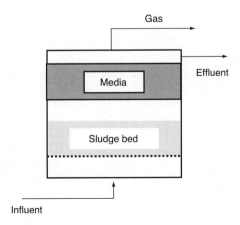

Fig. 9.9. Layout of a hybrid reactor

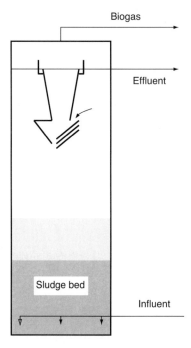

Fig. 9.10. Schematic diagram of an EGSB reactor

of the two types of digester. The EGSB reactors are taller, typically, having a height of 12–18 m compared with one of 5·5–6·5 m for UASB systems. Because of this and the loading rates used, this process takes up less space than a conventional UASB. The process uses recirculation to achieve the fluidizing velocities, and hydraulic retention times for full scale plants tend to be in the range of 3·5–15 h, although this, as well as the loading rate, will depend on the nature of the waste being treated and the standard of treatment required (Table 9.7).

Table 9.7. Examples of EGSB treatment plants

Waste	Volume (m^3)	Loading (kg/m^3 per day)	Flow (m^3/h)	COD concentration (mg/litre)
Sugar	767	21·3	165	4950
Potato	750	14·6	131	3500
Yeast	865	17·3	60	10 400
Brewery	780	29·5	300	2100
Chemical	275	10·0	6	20 000
Alcohol	250	11·2	21	5000
Fruit/vegetable	200	25·1	24	8800

Table 9.8. Performance of two EGSB plants treating potato wastewater (Zoutberg and Eker, 1999)

	Settled wastewater		Anaerobic effluent		Aerobic (final) effluent		
	COD	SS	COD	SS	COD	SS	BOD
A	4566	890	926	600	165	82	17
B	4500	1275	922	695	100	45	45

Once again, it must be remembered that if a very high-quality effluent is required, the digestion stage must be followed by an aerobic process. The need for doing this can be seen from an examination of Table 9.8, which shows the performance of two EGSB plants treating wastewaters from potato processing together with the final effluent quality after the aerobic polishing stage (Zoutberg and Eker, 1999).

Expanded and fluidized bed digesters

Both expanded and fluidized beds use a support medium which is usually sand (0·3–1 mm), but the differentiation between the systems is not clear-cut. The term expanded bed would probably be applied when the reactor was being operated at a bed expansion of 10–20%, while at bed expansions of greater than 30% it would be thought of as being fluidized. These expansions are achieved by the combined upflow of feed and recycled liquors (Fig. 9.11). The active biomass in expanded beds is the biofilm which colonizes the support medium (Fig. 9.12). A considerable amount of work has been done with

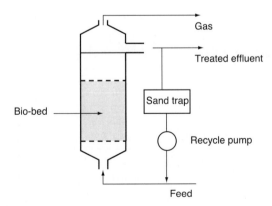

Fig. 9.11. Schematic diagram of a fluidized/expanded bed reactor

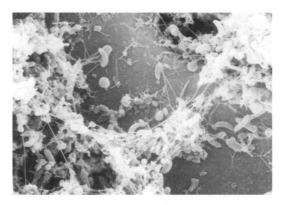

Fig. 9.12. Bacterial attachment to a sand particle from an anaerobic expanded bed reactor

laboratory and pilot-scale reactors (Table 9.9), but information about full-scale installations is less readily available. Expanded bed digesters are systems which are highly engineered but, despite this, some full-scale plants have been constructed. For example, an 80 m^3 digester in Australia has been built to treat the wastewater from a cereal processing plant with a loading rate of 7·5 kg BOD/m^3 per day (Barnes and Fitzgerald, 1987). Switzenbaum (1985) has also reported that an expanded bed digester treating a flow of 380 m^3/d from a soft drink bottling plant achieved a COD removal of 77% at a loading rate of 9·5 kg COD/m^3 per day.

Anaerobic baffled reactor

Although there are a number of variations in the way the anaerobic baffled reactor is assembled, the basic concept is that there should be a serpentine flow under and over baffles, as is shown in Fig. 9.13. Essentially, the variations relate to how the feed is introduced and where the gas is taken off (Barber and Stuckey, 1999). Hybrid

Table 9.9. Performance of anaerobic fluidized bed reactors

Waste	COD (g/litre)	Loading rate (kg COD/m^3 per day)	Removal (%)
Acid whey	50·3–56·1	13·4–37·6	72–83·6
Soft drinks	6	4–18·6	66–89
Whey permeate	27·3	5·3–7·4	82
Sweet whey	5–20	8·2–29·1	58·9–92·3

Anaerobic digestion of wastewaters

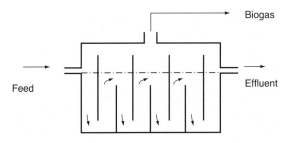

Fig. 9.13. Schematic diagram of an anaerobic baffled reactor

designs (see previous section) have also been suggested, with some or all of the compartments containing random or structured media. These various modifications have been mostly aimed at enhancing the solids' retention. The advantages claimed for this type of design include better resilience to shock loads, both hydraulic and organic, lower sludge yields and the ability partially to separate the various phases of the overall anaerobic process (Barber and Stuckey, 1999). Currently, the anaerobic baffled reactor is not common at full scale. However, one such plant was described in the review by Barber and Stuckey (1999). It was treating the flow from a population equivalent of 2500 and was operated with a liquid upflow velocity of 3 m/h. It achieved a COD removal of about 70% at loading rates varying between 0·4 and 2·0 kg/m^3 per day.

To digest or not to digest

Costs
Anaerobic digestion should not be thought of as a cheap option. Typically, in 2000 capital costs were of the order of £1000–£1200 per m^3. Expressed another way, capital costs range from 1 to 1·6 £/kg COD removal per annum and running costs from 6 to 15 pence/kg COD removed. These figures do not take into consideration secondary aerobic treatment or sludge handling/disposal.

Comparison with aerobic processes
One of the biggest advantages of anaerobic digestion is the low sludge yield in comparison with an aerobic option (Fig. 9.14). This will obviously affect operating costs which will include the costs of sludge handling and disposal. In addition, anaerobic processes do not need

Wastewater treatment and technology

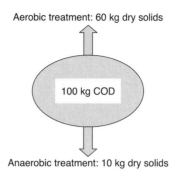

Fig. 9.14. A comparison of the sludge yield from aerobic and anaerobic processes

to be aerated and do generate an energy-rich by-product, the biogas (see Chapter 8). As such, they can be considered to be 'energy rich'. Anaerobic digesters can also tolerate shut down periods, something which would cause problems with aerobic reactors, and they will restart relatively quickly. This is a distinct bonus when the wastewaters are being generated by the processing of seasonal fruit and vegetables. Further advantages are listed in Table 9.10.

Basic considerations

There are a number of aspects which need to be taken into account when anaerobic digestion is being considered as a process option. Many industrial effluents may be nutritionally imbalanced with an excess of protein or carbohydrate. In these cases it will be necessary to add nutrients to achieve a balance for BOD:N:P of $100:0 \cdot 5:0 \cdot 1$.

Table 9.10. The advantages and disadvantages of anaerobic digestion

Advantages	Disadvantages
Running costs are low. There is no aeration and the mixing energy is low	High capital cost
	Heating may be required
Low sludge production	There may be odour problems with H_2S
Sealed, therefore, no aerosols	Only a pre-treatment
Can be a simple, flexible treatment with a low space requirement	Long initial start-up and recovery if the process is disrupted
By-products (gas and sludge) are useful	Pathogen reduction is not high
Restarting after seasonal shut-down is rapid	Additional alkalinity may be required
Anaerobic growth rate is low and, therefore, nutrient requirement is less	Nutrients are not removed completely

However, with industrial effluents the COD rather than the BOD is more likely to be the parameter used to measure the degree of organic pollution and, since the phosphorus requirement is of only minor significance, the COD:N ratio is, arguably, the critical consideration. The target value for this ratio is 400:7 (Henze and Harremoes, 1983). In addition to balancing the major nutrients, trace amounts of iron, cobalt and nickel may need to be added.

Toxicity is more likely to be a problem with industrial wastewaters than with waste sewage sludge. With food industry wastes, the chemicals used for cleaning are likely to have adverse effects on the digestion process. The concentrations of relatively simple chemicals may also be greater and may be sufficiently high to cause inhibition. Potassium and calcium, for example, can have moderately inhibitory effects at concentrations of 2·5–4·5 g/litre (McCarty, 1964). Treatability should, therefore, be assessed, and one way of doing this is to measure the specific methanogenic activity (litres of CH_4/g VSS per day) of an anaerobic sludge when fed with the wastewater under consideration and comparing the value with that obtained with some standard feed (Stuckey et al., 1980; James et al., 1990).

The other main points which need to be considered are

- Start-up, how to do it, the length of time it will take and what type of seed to use. There may also be a need to consider acclimatization of the biomass;
- Long-term performance and reliability, particularly in relation to fluctuations in the quality of the feed;
- Whether there is a need for re-circulation;
- Some wastewaters, dairy wastes for example, contain fats and oils. These can foam in gas-stirred digesters which are vigorously mixed and, therefore, precautions must be taken to ensure that gas lines and outlet pipes do not become blocked.

Having selected an anaerobic digestion option, it is important to ensure that proper control of the process is exercised. Conventional methods based on gas yield, alkalinity and VFA concentrations (Chapter 8) reflect what has already happened. However, the high-rate processes such as those being discussed in this chapter need the rapid detection and correction of process anomalies. Switzenbaum et al. (1990) have reviewed the measurement of a range of alternative parameters for assessing the state of an anaerobic digester. Some of these, such as the measurement of adenosine triphosphate or enzyme activity, which characterize the sludge, are not really amenable to rapid digester control because they would not be available as on-line methods. The realistic alternatives which could be used on-line are

- alkalinity (Hawkes et al., 1994, or the VFA:alkalinity ratio (Zickefoose and Hayes, 1976);
- hydrogen (Guwy et al., 1997);
- carbon monoxide (Hickey et al., 1987).

Modes of operation

When the anaerobic digestion of liquid wastes is being considered, the usual operating temperature is in the mesophilic range of 30–37 °C and the wastewater being treated is strong. Thus, all the data presented so far relates to these conditions. However, it is possible to consider anaerobic digestion for municipal sewage whose COD is appreciably lower than industrial wastewaters. In warm parts of the world it is possible to operate anaerobic digesters without using the energy in the biogas to provide heating. These two aspects, the treatment of domestic sewage and the use of ambient temperatures, have been brought together in a number of tropical countries using, mainly, UASB reactors. The performance of these digesters has been discussed in a review by Seghezzo et al. (1998). Some of the plants handle an appreciable flow. For example, the plant at Bucaramanga in Colombia was designed to treat $31\,000\,m^3/d$. Operating with a hydraulic retention time of five hours, it achieved a COD removal of 45–60% and an SS removal of about 60% (Schellinkhout and Osorio, 1994). In their review, Seghezzo et al. (1998) note that, in many cases, post treatment is needed to meet the standards required for discharge. They also note that, although treatment efficiency was not affected by low winter temperatures, biogas production decreased during the coldest period when the temperature reached 20 °C.

It is also possible to consider operation at other temperatures, the most usual being the thermophilic range of 50–60 °C. Some wastewaters are hot as they are generated and it is, therefore, logical to consider thermophilic anaerobic digestion. Doing this obviates the need for cooling. Operating in this temperature range has the advantage of being able to use a more compact reactor because of the greater activity of the biomass, up to 148 mmol methane/g VSS per day for thermophilic UASB granules compared to values of 10–42 mmol methane/g VSS per day for mesophilic granules (Schmidt and Ahring, 1996). However, although there are reports about the operation of some full-scale digesters (e.g. Vlissidis and Zouboulis, 1993), this variation of anaerobic technology is not widely applied on an industrial scale.

Prevalence of anaerobic digestion

Currently, there are some 900 full-scale plants in the world (Habets, 1996), with two thirds of these being UASB reactors. Within Europe, some 75% of the installed plant are UASBs, filters or contact digesters (Fig. 9.15). The majority of the wastes which are treated by this technology are from the agricultural/food industry (Fig. 9.16). However, non-food industry wastes can be and are treated by anaerobic digestion. The majority of these (c. 80%) are from the paper industry. The performance of one such plant is shown in Fig. 9.17.

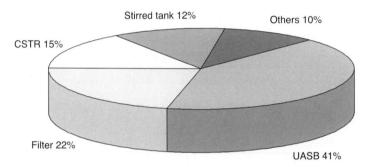

Fig. 9.15. Anaerobic digestion plants in Europe (Wheatley et al., 1997)

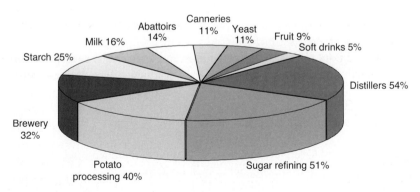

Fig. 9.16. Types of wastewater treated by anaerobic digestion (Wheatley et al., 1997)

Wastewater treatment and technology

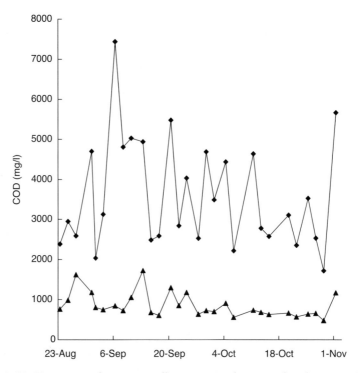

Fig. 9.17. Treatment of a paper mill wastewater by anaerobic digestion: COD values for influent (♦) and effluent (▲)

References

Barber, W. P. and Stuckey, D. C. (1999) The use of the anaerobic baffled reactor (ABR) for wastewater treatment: A review, *Water Res.*, **33**, 1559–78.

Barnes, D. and Fitzgerald, P. A. (1987) Anaerobic wastewater treatment processes, in *Environmental Biotechnology*, Forster, C. F. and Wase, D. A. J. (eds), Ellis Horwood, Chichester, pp 57–113.

Choi, E. and Rim, J. M. (1991) Competition and inhibition of sulfate reducers and methane producers in anaerobic treatment, *Wat. Sci. Technol.*, **23(7–9)**, 1259–64.

Guwy, A., Hawkes, F. R., Hawkes, D. L. and Rozzi, A. G. (1997) Hydrogen production in a high rate fluidized bed anaerobic digester, *Water Res.*, **31**, 1291–8.

Habets, L. (1996) Overview of industrial wastewater treatment, *Ind. Anaerobic Wastewater Treatment Conf.*, Sept., SCI, London.

Hawkes, F. R., Guwy, A. J., Hawkes, D. L. and Rozzi, A. G. (1994) On-line monitoring of anaerobic digestion: Application of a device for continuous measurement of bicarbonate alkalinity, *Wat. Sci. Technol.*, **30(12)**, 1–10.

Henze, M. and Harremoes, P. (1983) Anaerobic treatment of wastewater in fixed film reactors — a literature review, *Wat. Sci. Technol.*, **15**(8–9), 1–101.

Hickey, R. F., Vanderwielen, J. and Switzenbaum, M. S. (1987) Production of trace levels of carbon monoxide during methanogenesis of acetate and methanol, *Biotechnol. Lett.*, **9**, 63–6.

Hulshoff-Pol, L. W., Dolfing, J., Velzeboer, C. T. M. and Lettinga, G. (1982) Granulation in UASB reactor, *Paper presented at IAWPR Specialised Seminar on Anaerobic Treatment of Wastewater in Fixed Film Reactors*, Copenhagen, September.

James, A., Chermicharo, C. A. L. and Campos, C. M. M. (1990) The development of a new methodology for the assessment of specific methanogenic activity, *Water Res.*, **24**, 813–26.

McCarty, P. L. (1964) Anaerobic waste treatment fundamentals, part three, toxic materials and their control, *Public Works*, November, 91–4.

Morgan, J. W., Goodwin, J. A. S., Wase, D. A. J. and Forster, C. F. (1990) The effects of using various types of carbonaceous substrate on UASB granules and on reactor performance, *Biol. Wastes*, **34**, 55–71.

Morgan, J. W., Evison, L. M. and Forster, C. F. (1991) Upflow sludge blanket reactors: The effect of bio-supplements on performance and granulation, *J. Chem. Tech. Biotechnol.*, **52**, 243–55.

Patel, G. B. and Sprott, G. D. (1990) *Methanosaeta cocilii* gen., sp. nov. ('*Methanosaeta cocilii*') and *Methanosaeta thermoacetophilia* nom. rev., comb. nov., *J. System. Bact.*, **40**, 79–82.

Schellinkhout, A. and Osorio, E. (1994) Long-term experience with the UASB technology for sewage treatment on large scale, *Proc. 7th Int. Symp. on Anaerobic Digestion*, Cape Town, South Africa, pp 251–2.

Schmidt, J. E. and Ahring, B. K. (1996) Granular sludge formation in upflow anaerobic sludge blanket (UASB) reactors, *Biotech. Bioeng.*, **49**, 229–46.

Seghezzo, L., Zeeman, G., van Lier, J. B., Hamelers, H. V. M. and Lettinga, G. (1998) A review: the anaerobic treatment of sewage in UASB and EGSB reactors, *Bioresource Technol.*, **65**, 175–90.

Shore, M., Broughton, M. W. and Bumstead, N. (1984) Anaerobic treatment of wastewater in the sugar beet industry, *Wat. Pollut. Control*, **83**, 499–506.

Smith, M. O., Ferrall, J., Smith, A. J. T., Winstanley, C. I. and Wheatley, A. D. (1988) The economics of effluent treatment: a case study at Bovril, *Int. Biodeterioration*, **25**, 97–105.

Stuckey, D. C., Owen, W. F., McCarty, P. L. and Parkin, G. F. (1980) Anaerobic toxicity evaluation by batch and semi-continuous assays, *J. Wat. Pollut. Control Fed.*, **52**, 720–9.

Switzenbaum, M. S. (1985) Fluidised bed anaerobic reactors, in *Comprehensive Biotechnology Vol. 4*, Robinson, C. W. and Howell, J. A. (eds), Pergamon Press, Oxford, pp 1017–26.

Switzenbaum, M. S., Giraldo-Gomez, E. and Hickey, R. F. (1990) Monitoring of the anaerobic methane fermentation process, *Enz. Microb. Technol.*, **12**, 722–30.

Vlissidis, A. and Zouboulis, A. (1993) Thermophilic anaerobic digestion of alcohol distillery wastewater, *Bioresource Technol.*, **43**, 131–40.

Wheatley, A. D., Fisher, M. B. and Grobicki, A. M. W. (1997) Applications for anaerobic digestion for the treatment of industrial wastewater in Europe, *J. CIWEM*, **11**, 39–46.

Zickefoose, C. and Hayes, R. B. (1976) *Anaerobic Sludge Digestion: Operations Manual*, EPA 430/9-76-001.

Zoutberg, G. R. and Eker, Z. (1999) Anaerobic treatment of potato processing wastewater, *Wat. Sci. Technol.*, **40(1)**, 297–304.

10

Industrial wastewater treatment

Introduction

In considering industrial wastewater treatment, the options are

- provide full treatment and discharge to controlled waters. This will require a consent (Chapter 1);
- provide partial treatment and discharge to sewer. This will require a trade effluent consent (Chapter 1);
- provide no treatment and discharge to sewer. This will also require a trade effluent consent (Chapter 1).

If these options are to be assessed fully there will have to be a complete knowledge of the wastewaters being generated within an industrial site, particularly the flows from all the sub-units within the site and their composition. Also, when the components of the wastewaters are being examined, information about their biodegradability should be collated. If this information is not readily available, a comparison of the COD and BOD concentrations provides useful information.

A rigorous waste minimization programme may result in the reduction of the quantities of wastewater and, therefore, of the costs associated with each of these options. At the same time, checks should be carried out to ensure that accidental discharges to controlled waters, which could cause pollution, are minimized. In particular, all above ground storage tanks or drums should be located within a bund constructed of material impervious to the substance being stored (Fig. 10.1). The bunds are constructed so that, when the storage tanks are filled or material is taken out of them, any spillages or drips are contained within the bund. Also, because the factory owner is liable for acts of third parties,

Wastewater treatment and technology

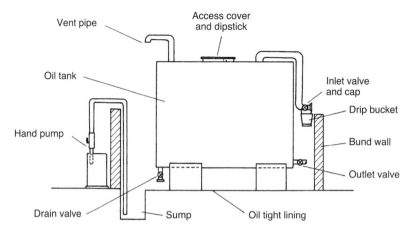

Fig. 10.1. Storage tank showing proper bunding facilities

all necessary precautions should be taken to ensure that intruders cannot cause discharges to occur from either storage tanks or any other part of the site. An integrated pollution control (IPC) study may be needed, which would also identify the best practical environmental option (BPEO).

Integrated pollution control

The concept of integrated pollution control (IPC) is, legally, associated with the Integrated Pollution and Control Directive (96/61/EEC) and, in the UK, the Environmental Protection Act, 1990. Integrated pollution control aims to restrict the releases of polluting material to air, water and land. Much of the previous legislation has only examined one of these sinks at a time. In the UK, the Act is operated by the Environment Agency for England and Wales, the Scottish Environmental Protection Agency and the Environment and Heritage Service for Northern Ireland, and aims to

- use the best available techniques not entailing excessive cost (BATNEEC);
- move away from the 'end of pipe' technology;
- restrict Red List material.

Thus, with IPC, the philosophy is to

- not move pollutants from one medium to another, i.e. from solution (water) to a solid which must be dumped (land) or burnt (air);

- limit and control the outputs;
- consider whether the type of the pollutant can be changed;
- develop training, management and education within the company.

Integrated pollution control applies to a range of prescribed activities, such as the chemical, energy, metals, oil and waste management industries, together with food production, the pulp and paper industry and some intensive livestock rearing (Webb, 1994). Guidance notes, which are aimed at assisting in the consistent determination of BATNEEC and BPEO for the specified industries, have been provided by the regulatory authorities.

When an IPC licence is issued, it covers emissions to air, water and land as well as noise emissions. The aim of the licence is to ensure that emissions of prescribed substances are prevented or, where that is not possible, minimized, and that pre-defined compounds which were emitted are rendered harmless by BATNEEC. The use of a single IPC licence for discharges and processes means that the discharges/processes will conform to National and International standards and practices. The licence also ensures consistency between emissions and allows for the monitoring and policing of discharges.

Industrial wastewaters

A variety of industrial wastewaters can be identified

- Readily biodegradable organic wastes such as those from the food and drinks industry;
- Organic wastes which may or may not be amenable to biological treatment; for example, paper, petrochemicals or textile wastes;
- Inorganic wastewaters such as plating wastes which require physico-chemical treatments.

Biological treatment

The biological treatment processes which are used will be those which have already been discussed: variants of the activated sludge process, biofilm reactors or anaerobic digestion. There may be differences in the loading rates and retention times and different pre-treatments may be used. These will tend to reflect the strength and nature of the wastes. It is also likely that a greater use will be made of flow-balancing tanks and the balancing of nutrients ($C:N:P$) needs to be watched very carefully. Some wastes will contain large amounts of

Wastewater treatment and technology

Table 10.1. Characteristics of trade wastes

Industry	BOD (mg/litre)	COD (mg/litre)	N (mg/litre)	P (mg/litre)	BOD:N:P
Beet factory	930	1600	16·4	3·4	274:4·8:1
Brewery	775	1220	19·2	7·6	102:2·5:1
Dairy	2300	4500	64	48	50:1·3:1
Thermomechanical pulping	370	3300	3·85	2·30	160:1·7:1
Molasses stillage	30 000	100 000	1500	150	200:10:1
Municipal sewage — settled	215	449	17	9·8	22:1·7:1

carbon with little N or P (Table 10.1). For aerobic treatment, the ideal figure for the C:N:P ratio is about 100:5:1. Anaerobic processes require fewer nutrients and have an optimum BOD:N:P ratio of 100:0·5:0·1 (Gray, 1999). An example of biological treatment using conventional aerobic processes is the use of high-rate filters (Chapter 3) for the treatment of a brewery wastewater (Fig. 10.2). The filters, which had a diameter of 21 m and a depth of 5·3 m, were operated with an irrigation rate of 1·1 m^3/m^2 per hour. The clarifiers, which were operated with an overflow rate of 1·6 m/h, were 18 m in diameter and had a side-wall depth of 2·5 m. Their performance is provided in Table 10.2. The activated sludge process (Chapter 4) is also used for the treatment of industrial wastewaters. The plant, whose performance is shown in Fig. 10.3, was operated with an organic loading rate of 0·16 kg BOD/m^3 per day. The data show that a conventional wastewater treatment plant can be used successfully with a strong and potentially variable influent. This is a confirmation of the survey by Saunamäki (1997), which showed that activated sludge plants treating paper mill wastewaters in

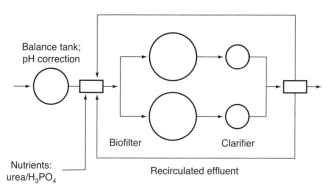

Fig. 10.2. Schematic diagram of a biological treatment plant for a brewery wastewater

Table 10.2. Performance of high-rate filters treating a brewery wastewater (Fig. 10.2)

	pH	SS (mg/litre)	BOD (mg/litre)
Raw wastewater	4–10	500	1100
Treated effluent	7–8	180	280

Finland were, in general, capable of achieving a BOD removal of more than 90% at loading rates of up to 0·4 kg BOD/m^3 per day.

Domestic sewage will contain fats, oils and greases (FOG), typically at concentrations of 40–100 mg/litre. Although this type of material is biodegradable, it can cause disruptions of biological processes. However, industrial effluents are of greater concern than domestic sewage. A wide range of manufacturing processes will generate wastewaters which contain appreciable amounts of FOG. These include the food industry, particularly milk processing, meat and fish processing and the production of edible oils, wool scouring, the petrochemical

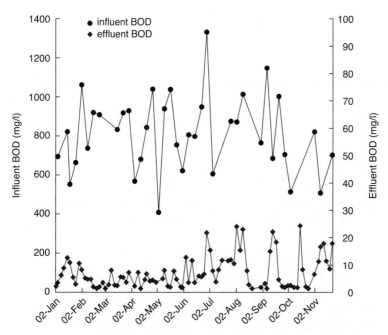

Fig. 10.3. Performance of an activated sludge plant treating wastewater from a paper mill

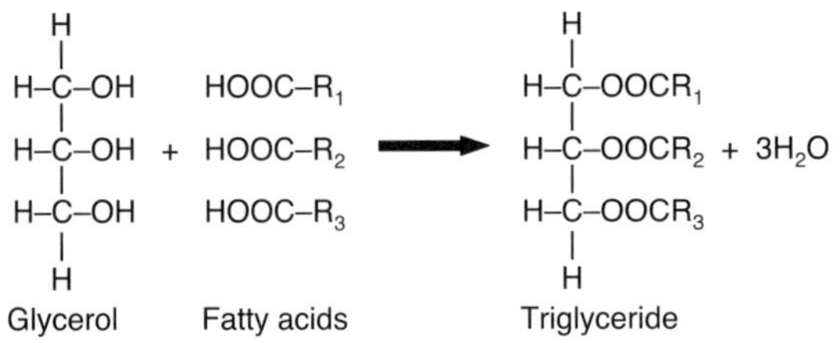

$$
\begin{array}{l}
\text{H} \\
| \\
\text{H–C–OH} \quad \text{HOOC–R}_1 \\
| \\
\text{H–C–OH} + \text{HOOC–R}_2 \\
| \\
\text{H–C–OH} \quad \text{HOOC–R}_3 \\
| \\
\text{H} \\
\text{Glycerol} \quad \text{Fatty acids}
\end{array}
\longrightarrow
\begin{array}{l}
\text{H} \\
| \\
\text{H–C–OOCR}_1 \\
| \\
\text{H–C–OOCR}_2 + 3\text{H}_2\text{O} \\
| \\
\text{H–C–OOCR}_3 \\
| \\
\text{H} \\
\text{Triglyceride}
\end{array}
$$

Fig. 10.4. *Triglyceride formation where R_1, R_2 and R_3 denote alkyl groups*

industry, and hotels and restaurants. For example, the FOG concentration in the effluent from a fish processing factory could be as high as 2880 mg/litre. The disruption of bio-processes is particularly true of anaerobic digestion where the long chain fatty acids, such as oleate, can be inhibitory (Hanaki et al., 1981; Angelidaki et al., 1990). Sayed et al. (1984) have also reported a high sludge yield during the treatment of slaughterhouse wastes. This was thought to be due to the accumulation of poorly degradable solids in the raw wastewater.

Fats, oils and greases (FOG) are a collective name for a wide range of compounds. They may be petroleum based or they may have a vegetable or animal origin. Petroleum-based oils will, in general, be long chain aliphatic compounds. The FOG which has a vegetable or animal origin will consist primarily of triglycerides. That is to say, a three-carbon glycerol molecule which has been esterified with three straight chain fatty acids (Fig. 10.4). Ninety-five per cent of most seed oils are made up of triglycerides (Sheppard et al., 1978). On a weight for weight basis, the fatty acids constitute 94–96% of the triglyceride molecule (Sonntag, 1979). The other molecules which may make up FOG include monoglycerides, diglycerides, free fatty acids, sterols, sterol esters, hydrocarbons, phosphatides, glycerol ethers and glycolipids.

The fatty acid molecule is made up of a long hydrocarbon chain, which is non-polar and hydrophobic, and a strongly polar carboxyl group which is hydrophilic. The fatty acids encountered in edible fats and oils contain 4–24 carbon atoms with 16-C and 18-C being the most common. The hydrocarbon chains in the fatty acids may be saturated or unsaturated (Table 10.3). The acids which exist in edible fats and oils vary over a wide range, and there are many published lists of the composition of these materials (Harwood and Geyer, 1964; Thomas and Paulicka, 1976). Table 10.4 lists some of these and uses the symbolism of CX:Z, where X denotes the number of carbon atoms and Z the number of double bonds. There will be other non-glyceride compounds present but, in refined vegetable oils, these will usually constitute less than 2%.

Table 10.3. Examples of saturated and unsaturated fatty acids present in edible fats and oils

Common name	Formula	Structure
Saturated fatty acids		
Palmitic acid	$C_{16}H_{32}O_2$	$CH_3(CH_2)_{14}COOH$
Stearic acid	$C_{18}H_{36}O_2$	$CH_3(CH_2)_{16}COOH$
Unsaturated fatty acids		
Oleic acid	$C_{18}H_{34}O_2$	$CH_3(CH_2)_7CH=CH(CH_2)_7COOH$
Linoleic acid	$C_{18}H_{32}O_2$	$CH_3(CH_2)_4CH=CHCH_2CH=CH(CH_2)_7COOH$

In the context of wastewater treatment, the analysis of FOG or of separable oil and grease is imprecise and, therefore, there will tend to be less than perfect knowledge about the lipid content of a wastewater. This means that the behaviour of these compounds in a wastewater treatment plant is not as well understood as it might be. The microbial

Table 10.4. Fatty acid composition of some edible fats and oils (Thomas and Paulicka, 1976)

Fatty acid	Chicken fat	Beef tallow	Mutton tallow	Corn oil	Olive oil	Sunflower oil
C20:4		0·4	0·4			
C20:2		–	–			
C20:1		–	–	0·1		
C20:0		0·1	–	–	0·9	0·4
C19:0		0·1	0·8	0·9	–	–
C18:3	1·3	0·6	1·3	57·0	0·6	0·5
C18:2	18·9	2·2	4·0	27·5	10·0	68·2
C18:1	41·6	38·7	33·3	2·2	71·1	18·6
C18:0	6·4	21·6	24·5	–	2·5	4·7
C17:1	0·1	0·7	0·5	–	–	–
C17:0	0·3	1·5	2·0	0·1	–	–
C16:1	6·5	3·4	2·5	12·2	1·2	0·1
C16:0	23·2	25·5	23·6	–	13·7	6·8
C15:1	–	0·2	0·3		–	–
C15:0	–	1·3	0·8			–
C14:1	0·2	0·2	0·3			–
C14:0	1·3	3·3	5·2			0·2
C13:0	–	–	–			–
C12:0	0·2	0·1	0·3			0·5
C11:0		0·1	0·2			

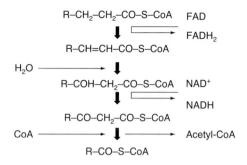

Fig. 10.5. β-oxidation pathway for fatty acids

degradation pathways of pure lipid compounds are known. Triglycerols and phospholipids are broken down to fatty acids by a wide variety of micro-organisms. These are then broken down further by what is known as β-oxidation. As is shown in Fig. 10.5, the fatty acid molecule, after being converted to a fatty acid CoA ester, is degraded in a sequence of four reactions to an ester which has been shortened by two carbon units. The sequence shown in Fig. 10.5 will continue until it can proceed no further. In the case of an odd-chain-length acid, the final products will be acetyl-CoA and propionyl-CoA. An even-chain-length acid will produce only acetyl-CoA. Alkanes are degraded through the reactions shown in Fig. 10.6 and the resultant fatty acids then undergo β-oxidation.

Most of the FOG associated with the industries mentioned earlier will be naturally occurring and, as such, will be biodegradable, but there will be a proportion which is synthetic or which has been chemically modified. This latter group may not be as readily degradable. It has been estimated that some 20% of the oils and fats used by industry fall into this group and represent some 10 million tonnes per

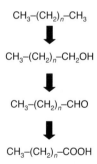

Fig. 10.6. Oxidation pathway for alkanes

year (Ratledge, 1992). It is recognized that only a very small proportion of this will end up in wastewater but it is, nevertheless, an area of concern.

In the UK, industrial wastewaters require a consent before they can be discharged to the sewerage system (Chapter 1) and, since they are not desirable in sewers the concentrations of FOG which are permitted tend to be restricted. A typical trade effluent consent would limit the concentration to about 100 mg/litre. This means that some form of on-site physico-chemical pre-treatment may be required. At its simplest, this would be a fat-trap or an oil interceptor (Capps et al., 1993). These are effective processes but they can suffer from overloading and poor maintenance (Hrudey, 1984). Essentially, there are two basic types of interceptor, the American Petroleum Institute (API) separator and the corrugated plate interceptor. The API separator removes free oil and insoluble solids, and consists of a rectangular basin through which the wastewater flows horizontally. Free oil rises to the surface because of the difference in specific gravities. The rate of separation is governed by Stoke's Law so that the rise velocity of an oil droplet, V (cm/s), is given by

$$V = \frac{[(2gr^2)(d_1 - d_2)]}{9\mu} \qquad (10.1)$$

where g = acceleration due to gravity (cm/s^2), r = equivalent radius of the oil droplet (cm), d_1 = density of the droplet (g/cm^3), d_2 = density of the water (g/cm^3), μ = fluid viscosity. In operating an API separator it is essential that there is no short-circuiting and that laminar flow conditions should prevail (Reynolds Number <500).

Corrugated plate interceptors are also based on Stoke's Law and use packs of inclined (45–60°) plates in the basin. These are mounted parallel to each other with a spacing of 2 to 4 cm (Fig. 10.7). As the oil droplets float upwards they are trapped in the corrugations of the plates and coalesce into larger globules. They then move up the surface of the plates to form a floating layer which can be skimmed off.

Flotation, using dissolved air (Chapter 8) (Arnold et al., 1995) or electrolytically generated gas (Lewin and Forster, 1974), is an alternative which has been successful, removing up to 90% of the applied BOD as fat. However, it is a process which can be overwhelmed by shock overloads (Sparshott, 1991). It has been claimed that the recovered fat and, in some cases, protein could be re-used (Iggleden and van Staa, 1984; Ziminska, 1985), but there is no real evidence that this is a marketable product.

Wastewater treatment and technology

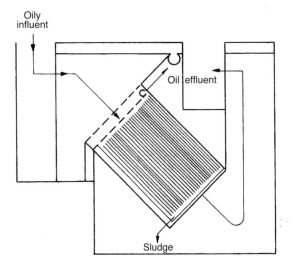

Fig. 10.7. Schematic diagram of a lamella plate interceptor

Enhanced biological treatment

The Municipal Wastewater Treatment Directive requirement for a COD limit (Chapter 1) has focused attention on non-degradable COD. In the case of wastes which contain recalcitrant organics, known as hard COD, some thought must be given to the removal of this fraction. Some of the compounds which constitute hard COD may be harmful to the environment as well as being non-biodegradable, for example, dye residues, particularly those from reactive and acid dyes, or halogenated organics. The former can impart colour to the receiving waters (Correia *et al.*, 1994), while the latter may be both persistent and toxic. The impact of hard COD may be reduced by segregating waste streams within the factory or by pre-treatment. The use of dissolved air flotation has already been mentioned as a pre-treatment to reduce FOG concentrations, but it is also one which finds considerable application in the treatment of other types of industrial wastes. It is used in the petroleum industry to remove oily residues, dyehouses to remove precipitated dyestuffs and poultry processing to remove blood and fat residues. However, it may be necessary to enhance the bio-process itself. One way of doing this is to use powdered activated carbon (PAC) in conjunction with the activated sludge process (O'Brien, 1992; Wong *et al.*, 1992; Costa and Marquez, 1998). There are several advantages of using PAC in this way. It will adsorb toxic organics which could affect both the process and the quality of the

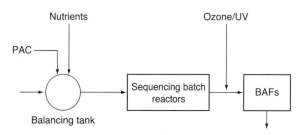

Fig. 10.8. *Schematic diagram of an enhanced biological treatment plant for an industrial leachate*

final effluent (Narbaitz et al., 1997). The PAC will also remove higher molecular weight organics, which are not easily biodegradable, by adsorption. The activated sludge will then be able to biodegrade lower molecular weight organics, which are not easily adsorbed by PAC. In addition, the PAC will act as a carrier for micro-organisms. This allows the specialist bacteria required to biodegrade intractable organics to accumulate. The presence of the PAC increases the sludge age of the organic component of activated sludge. This, in turn, allows a longer period for biodegradation to occur. The PAC may reduce the sludge volume index by up to 50%. This could help to increase the hydraulic retention time of the system without loss of solids from the clarifier. A further benefit is that PAC laden activated sludge de-waters more effectively than conventional activated sludge. This approach was used for the treatment of a flow ($500 \, m^3/d$) of industrial leachate and liquors from a groundwater remediation site. Two biological processes were used: a sequencing batch reactor, with a hydraulic retention time of 67 h, followed by a biological aerated filter with a hydraulic retention time of 3 h (Chapter 5). Powdered activated carbon and nutrients were added to the balancing tank (Fig. 10.8). The performance of the plant is given in Table 10.5.

Ozone (Chapter 7) can also be used to remove refractory COD prior to biological treatment (Baig and Liechti, 2001; Alvares et al., 2001). Most of the work using this approach has been done at laboratory or pilot-scale level, and has shown that the required dose for a maximum improvement in biodegradability would be between 0·23 and 1·04 mg O_3/mg COD (Alvares et al., 2001). However, the full-scale use of ozone as a precursor to biological treatment has been applied at an industrial chemical complex in Germany (Rice, 1997). Ozone has also been used after biological treatments to remove residual refractory compounds, particularly those producing colour (Churchley, 1998).

Table 10.5. Performance of wastewater treatment plant using PAC (Fig. 10.3)

	pH	COD (mg/litre)	BOD (mg/litre)	Amm.N (mg/litre)
Raw wastewater	6–8	4000	150	900
After SBRs	7–8	2800	15	20
After BAFs	7–8	250	5	2

Physico-chemical treatment

There are a range of advantages and disadvantages, in relation to biological processes, of using physico-chemical treatment. The advantages include the fact that they can be readily controlled and can be started and stopped rapidly without any adverse effects. In addition, they can accommodate a wide variation in flows and loads, and can be used on biologically resistant materials. The main disadvantages are that relatively large quantities of sludge may be produced, that high concentrations of dissolved salts may be present in the treated water, and operating costs may be higher.

When inorganic wastewaters and, in some cases, organic wastes are being treated physico-chemically, the technologies used can include

- precipitation/coagulation;
- oxidation/reduction;
- adsorption;
- membranes.

Treatments with coagulants

The use of coagulants (Table 10.6) is frequently preceded by the precipitation of soluble material. The chemistry involved may be

Table 10.6. Examples of precipitation/coagulation treatment

Industry	Treatment	Performance
Textile	Fe/Al salts at pH = 6–8	80–90% SS removal
Pulp/paper	Fe/Al salts at pH = 7–9	40–60% COD removal
		80–90% colour removal
Heavy metals	Lime, NaOH to pH = 8–11	<1 mg/litre residual soluble metal
Fluorides	$CaCl_2$ + Al salt pH = 6	<30 mg/litre residual fluoride
Paints	Fe/Al salts at pH = 7–9	>90% SS removal
		<80% COD removal

Industrial wastewater treatment

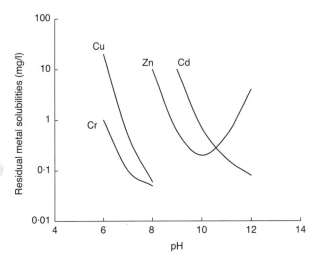

Fig. 10.9. Residual solubilities of metal ions following precipitation with calcium hydroxide

quite complex and may need to be optimized by pH correction (Fig. 10.9). Also, different precipitants can give different residual concentrations. For example, lime tends, in general, to give a better performance than sodium hydroxide for the precipitation of metals. Although, in the main, the chemicals used for precipitation reactions could be described as simple chemicals, metal ions or hydroxides, in specific cases where ultra-low residual concentrations are required, more complex chemicals may need to be used. An example of this is the ultimate treatment of wastewaters containing soluble mercury or cadmium which are both 'Red List' compounds (Chapter 1). These metals can be precipitated as insoluble sulphides, but this can result in there being excesses of sulphide in the treated liquors. Also, the precipitates can be difficult to de-water. Commercial, more complex alternatives such as trimercaptotriazine are also used. However, recent work has indicated that the metal compounds formed with these alternatives may be unstable (Matlock et al., 2001, 2002).

The coagulation of precipitated solids or wastewater solids usually involves the use of calcium, iron or aluminium salts, and their action can be enhanced by the use of polyelectrolytes such as polyacrylamides. Typical examples of wastewater treatment by this means are shown in Table 10.6. The action of coagulants can also be optimized by the use of the correct pH range.

Wastewater treatment and technology

Table 10.7. Examples of wastewater treatment by chemical oxidation

Pollutant	Oxidant	Performance
Cyanide	NaOCl at pH = 8·5–10	0 mg/litre CN
Phenolics	$H_2O_2 + Fe^{2+}$	<10 mg/litre phenol
Sulphides	Air + Mn^{2+}	<20 mg/litre S
Ammonia/amines	NaOCl at pH = 6–8	Complete destruction

Oxidation/reduction

A range of oxidation processes are available for wastewater treatment. These include the use of chlorine, ozone, hydrogen peroxide and potassium permanganate (Table 10.7). Oxidation can be used to effect the destruction of recalcitrant or problematic components in a wastewater, phenols for example, or to convert this type of material into one which is more amenable to bio-degradation. Reduction is not widely used. However, these two technologies are combined in the treatment of plating wastes which contain cyanides, chromates and residual heavy metals, with chlorine being used as the oxidant and sulphur dioxide as the reducing agent (Fig. 10.10). The chemistry involved is

$$HCN + HOCl = CNCl + H_2O$$

$$CNCl + H_2O = CNO^- + HCl + H^+$$

$$2CNO^- + 2HOCl = N_2 + CO_2 + 2HCl$$

$$Cr(VI) + SO_2 \Rightarrow Cr(III)$$

$$Cr(III) + \text{other metals} + OH^- \Rightarrow \text{insoluble hydroxides}$$

Adsorption

Activated carbon is arguably the most common and the most effective adsorbent for organics (Table 10.8). It is best used as a column of granular activated carbon (GAC) (Walker and Weatherly, 2000; Rajagopal and Kapoor, 2001). The GAC can be re-generated when, or just before, it reaches its full adsorptive capacity. This is done thermally, for example, by passing live steam through the bed. There will be some loss by irreversible adsorption and attrition. This is likely to be about 5–10% and will need to be made up by the addition of fresh carbon. Adsorption by GAC can also be used in conjunction with other treatment processes. A typical example of this is its use with secondary treatment and ozonation to produce water for industrial re-use (Bergna et al., 1999).

Industrial wastewater treatment

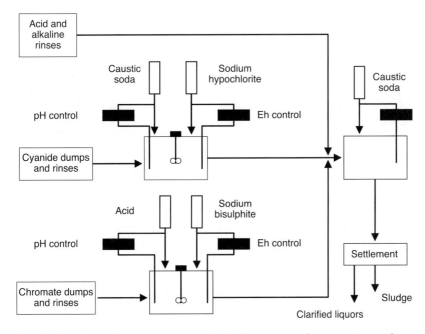

Fig. 10.10. Schematic diagram of a chemical treatment plant treating a plating wastewater

The use of GAC columns will also enable heavy metals to be removed from solution and because GAC is such an effective adsorbent, its performance in removing metal ions has been examined widely. For example, Bowers and Huang (1980) have investigated the removal of chromium (VI) from industrial wastewaters, and Wilczak and Keinath (1993) have evaluated the sorption of copper and lead. When this option is being used there is a need to be able to desorb the metal ions so that they can be recovered and the GAC re-generated. The study by Wilczak and Keinath (1993) also examined the desorption

Table 10.8. Organic compounds amenable to adsorption on activated carbon

Organic compounds	
Alkyl benzene sulphonates	Phenol, cresol, resorcinol
Gasoline, kerosene, oil	Textile dyes
CCl_4, perchloroethylene	Benzene, toluene, xylene
Naphthalene, biphenyls	Chlorobenzene toxaphene, DDT
Aniline, toluene diamine	Aldrin, eldrin

kinetics and Reed et al. (1995) evaluated a range of re-generation schemes for a GAC/lead system. Taken overall the performances imply that, all other factors being equal, GAC should be considered the most appropriate choice. However, the cost of GAC is such that it may not be a viable option in some parts of the world. Because of this, there has been considerable interest in the suitability of other sorbents with lower costs (Wase and Forster, 1997). Although there has been a lot of interest in the use of peat (Allen et al., 1992), waste biological material such as rice hulls and coir have been examined, either in their natural state or after carbonization (Wase and Forster, 1997). The use of immobilized bacteria and algae has also been assessed and processes based on this approach are available commercially. Two examples are AlgaSORB™, which is based on *Chlorella* immobilized in silica or polyacrylamide gels, and the bacteria-based (*Bacillus subtilis*) AMT-Bioclaim process (Garnham, 1997).

Whichever type of adsorption is used, the process is usually quantified in terms of adsorption isotherms, the two most common being the Langmuir and the Freundlich models. The former assumes the adsorbed species interact only with the sorption site and not with each other, that the adsorption energy of the sites is identical and does not depend on the presence of adsorbed species and that adsorption is limited to a monolayer. The model is described by

$$\frac{C_e}{q_e} = \frac{1}{K_L} + \left(\frac{a_L}{K_L}\right)C_e \qquad (10.2)$$

where q_e is the uptake of metal at equilibrium (mg/g), C_e is the concentration of metal species in solution at equilibrium (mg/litre), a_L and K_L are the Langmuir constants. The maximum metal uptake can then be derived as $Q_o = K_L/a_L$.

The Freundlich model, which is empirical, can be described by

$$\ln q_e = \ln K + \frac{1}{n} \ln C_e \qquad (10.3)$$

K (mg/g) provides a measure of the adsorption capacity and $1/n$ a measure of the sorption intensity.

While the Langmuir and the Freundlich models will provide information about the degree of adsorption which can be expected for a particular sorbent/sorbate system, the values for Q_o and K should not be used in isolation as they will depend on the conditions used for their derivation. For example, the initial pH will affect their value. Also, when the design of continuous systems is being undertaken an additional range of information, such as the length of unused bed, the bed depth service time and the empty bed residence time, is needed. These points have been discussed in detail by McCay and Allen (1997). These workers also describe the range of reactor types which can be used, including downflow contactors, pulsed bed systems and fluidized bed sorbers.

At the time of writing, GAC is most likely to be the first option examined for any adsorption system, and its cost will be an important aspect of any appraisal and depend on the particular application.

Enhanced oxidation

It is usual to differentiate between chemical oxidation and advanced oxidation.

Chemical oxidation uses hydrogen peroxide, chlorine, chlorine dioxide or potassium permanganate. At least one company in the USA offers a process based on permanganate for the oxidation of organics such as phenols, acrolein and olefins.

Advanced oxidation uses a variety of techniques to generate hydroxyl radicals (Munter et al., 2001; Perkowski et al., 2000). These are non-specific oxidants which are ideal for converting recalcitrant organics to carbon dioxide and water.

Hydroxyl radicals (OH^*) may be generated in several ways:

(a) Fenton's reagent (H_2O_2 and Fe^{++})

$$H_2O_2 + Fe^{2+} = Fe^{3+} + OH^- + OH^*$$

This has been used to remove phenolics and other aromatics (Rivas et al., 2001). The optimum pH is 3–4 and, typically, reaction times are 30–60 minutes.

(b) UV and hydrogen peroxide. This uses a wavelength of less than 400 nm and gives a very effective use of the peroxide

$$H_2O_2 + energy = 2OH^*$$

(c) UV and ozone

$$H_2O_2 + O_3 = 3O_2 + 2OH^*$$

(d) Catalytic enhanced destruction of organic contaminants using titanium dioxide or ultrasound as a catalyst to enhance the effect of hydroxyl radicals.

Combinations of ultraviolet radiation, ozone, hydrogen peroxide and catalysts have been used with a variety of recalcitrant pollutants. These

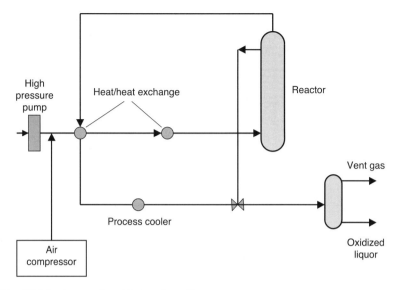

Fig. 10.11. A wet air oxidation flow diagram

include atrazine (Prado and Esplugas, 1999), the wastewater from polyester resin production (Bertanza et al., 2001), high explosive contaminated wastewater (Alnaizy and Akgerman, 1999) and pharmaceutical wastewater (Hoefl et al., 1997).

Wet air oxidation (WAO) is another form of advanced oxidation (Fig. 10.11). It is a technology which uses a combination of air or oxygen, high pressure (50–200 bar) and high temperature (200–325 °C). At these temperatures and pressures the solubility of oxygen is greatly increased so that there is a strong driving force for the oxidation of any organics which are present, as can be seen in Table 10.9. Typical retention times are 45–60 minutes. The oxidation generates heat and it is claimed that the process is self-sustaining if wastes with a high COD are treated. High is generally taken as being more than 20 g/litre. Wet air oxidation is most appropriate for waste streams which contain dissolved or suspended organics with concentrations of more than 500 mg/litre. The type of wastes which have been treated by WAO include paper mill black liquors, trinitrotoluene red water and phenolics (Maloney et al., 1994). Wet air oxidation has also been used to treat the wastewater from a pharmaceutical plant which contained phenols, p-aminophenol and reduced sulphur compounds. The plant, which was designed to treat a flow of 112 litres/min, was operated at 260 °C and a pressure of 100 bar. The process reduced the COD concentration from 69 100 mg/litre to 2590 mg/litre and the

Table 10.9. Performance for a WAO plant operating at 210 °C and 4 MPa (40 bar)

Material	Influent (mg/litre)	Effluent (mg/litre)
Trimethyl benzene	22	0·015
1-methyl naphthalene	9·6	BDL
Naphthalene	0·68	0·0025
Bis(2-ethylhexyl)phthalate	34	0·0089
Trichloroethylene	38	BDL
1,1,2-trichloroethane	550	BDL

BDL = below detection limit

p-aminophenol concentration from 6950 mg/litre to less than 0·1 mg/litre. Most of the reduced sulphur compounds were also reduced, resulting in a sulphate concentration in the treated effluent of some 49 000 mg/litre.

The LOPROX process (LOw PRessure wet OXidation) is a catalysed system which operates in an acidic environment at a lower temperature and pressure than normal wet oxidation processes (Hug et al., 1998). Typically the temperature would be less than 200 °C and would use a pressure of 5–20 bar. The catalyst is iron-based and could be used for wastes with a COD of 10–20 g/litre. The process can be used in two ways: to produce a complete oxidation of the organic contaminants such that the COD is reduced by more than 90%, or to achieve a partial COD reduction and to oxidize the residual matter to such an extent that it is highly biodegradable.

These techniques are not cheap and consideration must be given to where they are applied. The volumes involved imply that they must be used at the factory site.

Recovery

One way of dealing with refractory material is to move onto the factory site and practice waste minimization or materials recovery. Indeed, recovery may be thought of as a form of treatment. Recovered material may be raw material or product, and the quality, quantity and cost of the recovered material will be critical in judging the feasibility of any recovery exercise. For organic compounds with a COD, recovery probably means adsorption, extraction or the use of membranes.

Table 10.10. Comparison of membrane processes

Membrane process	Pore size (μm)	Molecular weight cut-off (daltons)	Operating pressure (bar)
Microfiltration	<20 000	–	0·5–1·0
Ultrafiltration	0·01	100 000–500 000	0·3–7
Nanofiltration	0·001	400–1000	5–10
Reverse osmosis	<0·0001	<400	10–100

The use of GAC and biosorbents and their potential role for recovery of metals has already been discussed.

Membrane processes rely on the permeation of material through a porous medium, with the driving force, in most cases, being pressure. They can be thought of as processes which 'strain out' particles, molecules or ions, depending on the type of membrane being used and the pressure drop required (Table 10.10). The polymers used for the membranes include polysulphones, cellulose acetate or polyamides. They are very thin, typically less than 1·0 μm, and must be supported on a stronger, highly porous frame. These processes need a large area of membrane to achieve a reasonable throughput. The configuration of the membrane is, therefore, important if a compact plant is to be designed (Table 10.11).

Although membrane systems can be used for the removal of polluting matter, in the appropriate circumstance, removal can be equated to recovery, and there are a variety of membrane processes which can

Table 10.11. Membrane configurations

Form	Cost	Advantages	Disadvantages	Applications
Pleated cartridge	Low	Cheap, robust, compact	Easily fouled. Cannot be cleaned	MF
Plate and frame	High	Can be dismantled for cleaning	Cannot backflush	UF RO
Spiral	Low	Simple, robust, compact	Not easily cleaned. Cannot backflush	UF RO
Tubular	Very high	Easily cleaned. Tolerates high TSS	High capital and membrane costs	UF
Hollow fibre	Very low	Simple and compact	Sensitive to pressure shocks. High pressure for backflushing	RO MF

Industrial wastewater treatment

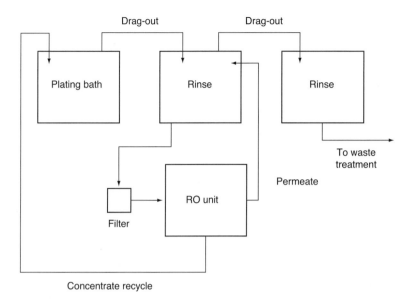

Fig. 10.12. *Schematic diagram of a reverse osmosis treatment plant for the recovery of water from plating rinsewater*

be applied to the removal/recovery of both inorganic and organic material. Indeed, reverse osmosis has been used in the paper/pulp industry for many years, mainly for the concentration of spent sulphite liquors. Similarly, ultrafiltration membranes have been used to recover protein from potato starch effluents or to treat wool scour effluents. The main problem with most membranes is the low flux (volumetric flow/ unit area in a given time), which means that very large areas are needed. Also membranes can foul. Certainly work is focusing on extending membrane fluxes but, as yet, this is not being used at a commercial level. Membrane systems can also be used for the concentration of metal ions, a form of recovery, in, for example, the nickel plating industry. The system shown in Fig. 10.12 used a cellulose acetate membrane with a flux of 19–27 litres/m^2 per hour and a pressure of 4000 kPa. The reverse osmosis unit received a flow of 0·227 m^3/h containing a total dissolved solids' (TDS) concentration of 4300 mg/litre, and produced a permeate with a TDS concentration of 86 mg/litre. The concentrate which was recycled to the plating bath contained a TDS concentration of 84 000 mg/litre.

Waste streams containing metals can also be treated with membranes. One such technology is microfiltration which will remove particles down to about 0·5 μm. Microfilters can be made from metal, ceramic or polymeric material. One such system is the Renovexx, a cross-flow filter which uses porous polyester tubes about 25 mm in diameter. As a suspension is pumped into the tubes, deposition of solids on the internal walls forms a dynamic membrane which enables the high quality of performance to be achieved. During the initial phase, there is a need for re-circulation to enable this 'membrane' to form. Eventually, the rate of filtration will decrease to an unacceptable level and at this point the tubes require cleaning. This is done by reducing the internal pressure and dislodging the accumulated solids by agitation. This approach was used to treat a wastewater flow of $20\,m^3/h$, which contained a suspended solids' concentration of 350 mg/litre including 50 mg Pb/litre. The effluent contained less than 0·1 mg Pb/litre.

Agricultural industry

The agricultural industry must be taken to encompass the production of livestock, cereals, fruit and vegetables, together with the manufacture of agrochemicals such as fertilizers and biocides. Taken overall, this can be considered as a major international industry, and the wastes and residues it produces have the potential to cause significant environmental impacts (Table 10.12 and Fig. 10.13).

Perhaps the most direct impacts of farming and farming practice are on the quality of raw water and, subsequently, on the quality of drinking water. The direct effects on raw water are organic contamination, either by silage liquors or organic slurries, both of which can be highly polluting and can, in extreme cases, cause major fish kills (Figs 10.14 and 10.15). Even if the degree of pollution is not sufficient to cause fish kills, there are likely to be heavy growths of 'sewage fungus' in the watercourse. This is not a pure fungus in the biological sense, but is a mixture of filamentous bacteria, such as *Sphaerotilus natans*, protozoa

Table 10.12. Strengths of wastes from agriculture (MAFF, 1998)

Material	BOD (mg/litre)
Silage liquors	30 000–80 000
Pig slurry	20 000–30 000
Cattle slurry	10 000–20 000
Drainage from slurry storage	1000–12 000
Raw municipal sewage	200–500

Industrial wastewater treatment

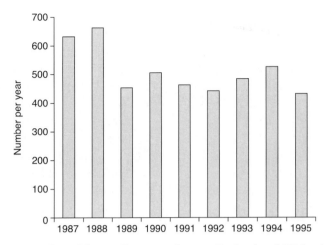

Fig. 10.13. *Number of farm pollution incidents in England and Wales during the period 1987–1995 (Lennox et al., 1998)*

and, occasionally, some fungi that grows as long grey strands attached to stones and aquatic plants in the watercourse, and is highly indicative of organic pollution (Webb, 1985).

Silage is produced by the fermentation of sugars in grass by the lactobacilli under anaerobic conditions. This reduces the pH to about 4·5 and preserves the grass. During the fermentation, liquors are generated. The amount produced can be up to 0·45 m^3/tonne of silage, depending on the dry matter in the crop, and will contain, for

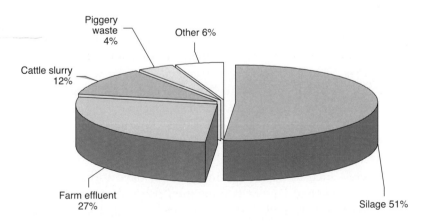

Fig. 10.14. *Fish kills in Northern Ireland in the period 1987–1995 (Lennox et al., 1998)*

Fig. 10.15. A fish kill (courtesy of the Environment Agency)

example, lactic acid, acetic acid and butyric acid. These liquors are highly polluting with BOD values of around 50 000 mg/litre (Table 10.10). As can be seen from Fig. 10.14, silage liquors are responsible for pollution incidents which were significant enough to cause fish kills. Legally, pollution of watercourses by silage liquors is controlled by the 1991 Control of Pollution (Silage, Slurry and Agricultural Fuel Oil) Regulations. In practice, if the grass is allowed to wilt before it is ensiled, silage effluent can largely be eliminated (Mason, 1988). Wilting reduces the dry matter from 14–17% in the uncut grass to a value of about 25%. Pollution control is, therefore, a combination of good practice and common sense.

The intensification of livestock rearing has meant that larger quantities of animal waste are generated in centralized locations (Hooda *et al.*, 2000). It has been estimated that the waste produced by the UK livestock industry (cattle, pigs and poultry) amounts to about 34 000 tonnes of dry solids per day (Dagnall, 1995). These are highly polluting; for example, cattle slurry can have BOD values of up to 20 000 mg/litre (Table 10.10). Furthermore, depending on the diet the animals have received, their excreted material may contain heavy metals such as copper or zinc. Slurries have to be stored and there can be polluted liquors draining from these storage areas. These liquors, together with milk parlour washings and drainage from fouled yards, can be considered as a potential point source pollution (Brewer *et al.*, 1999; Hooda *et al.*, 2000). These wastes could be treated, if the desire were there, by either aerobic or anaerobic processes (Miner et al., 2000).

Slurried animal wastes can also be treated by anaerobic digestion (Chapters 8 and 9) (Hawkes et al., 1985; Misi and Forster, 2002), but they do require considerable dilution before this can be done. Also, poultry manure can generate sufficiently high concentrations of ammoniacal-nitrogen during the digestion for it to become inhibitory (Webb and Hawkes, 1985).

Slurried animal wastes can also be treated by anaerobic digestion, but they do require considerable dilution before this can be done. Also, poultry manure can generate sufficiently high concentrations of ammoniacal-nitrogen during the digestion for it to become inhibitory. Further considerations are: can the farmer be bothered to manage a digester? What about the safety aspects of handling the biogas? Simplified digesters have been considered but they are not in widespread use in the UK. Centralized co-digestion of biowastes could be an alternative (Chapter 11).

Fish farming can also create problems in watercourses downstream of the farming zone. These are due to residual fish food, faecal matter and disease resulting from the intensive culturing of the fish. As well as imposing an input of organic contaminant, fish farms can also add nitrogen and phosphorus to downstream waters.

In those parts of the country where sheep are the main farming concern, pollution of watercourses and ground water can occur from the residues from sheep dipping (Virtue and Clayton, 1997). This is carried out to control parasites, with the active ingredients of the dip often being organophosphorus compounds. The contamination of watercourses by these chemicals can reduce the stream biota and, in extreme cases, cause fish kills (Hooda et al., 2000). The safe disposal of dip residues is, therefore, critical to maintaining good quality water. Soakaways, which have been used in the past, are now considered to represent a high risk to surface and ground waters. It is currently recommended that spent dip liquors should be diluted with three parts slurry or water and spread on land away from drains and streams (MAFF, 1998).

References

Allen, S., Brown, P., McCay, G. and Flynn, O. (1992) An evaluation of single resistance transfer models in the sorption of metal ions by peat, *J. Chem. Tech. Biotech.*, **54**, 271–5.

Alnaizy, R. and Akgerman, A. (1999) Oxidative treatment of high explosives contaminated wastewater, *Water Res.*, **33**, 2021–30.

Alvares, A. B. C., Diaper, C. and Parsons, S. A. (2001) Partial oxidation by ozone to remove recalcitrance from wastewaters — A review, *Environ. Technol.*, **22**, 409–28.

Angelidaki, I., Petersen, S. P. and Ahring, B. K. (1990) Effects of lipids on thermophilic digestion and reduction of lipid inhibition upon addition of bentonite, *Appl. Microbiol. Biotechnol.*, **33**, 469–72.

Arnold, S. R., Grubb, T. P., Thomas, P. and Harvey, P. J. (1995) Recent applications of dissolved air flotation, pilot studies and full scale design, *Wat. Sci. Technol.*, **31(3–4)**, 327–40.

Baig, S. and Liechti, P. A. (2001) Ozone treatment for biorefractory COD removal, *Wat. Sci. Technol.*, **43(2)**, 197–204.

Bergna, G., Bianchi, R. and Malpei, F. (1999) GAC adsorption of ozonated secondary textile effluent for industrial reuse, *Wat. Sci. Technol.*, **40(4–5)**, 435–42.

Bertanza, G., Collivignarelli, C. and Pedrazzani, R. (2001) The role of chemical oxidation in combined chemical, physical and biological processes: Experiences of industrial wastewater treatment, *Wat. Sci. Technol.*, **44(5)**, 109–16.

Bowers, A. R. and Huang, C. P. (1980) Activated carbon process for the treatment of chromium (VI) containing industrial wastewaters, *Wat. Sci. Technol.*, **13(1)**, 629–50.

Brewer, A. J., Cumby, T. R. and Dimmock, S. J. (1999) Dirty water from dairy farms, II: treatment and disposal options, *Bioresource Technol.*, **67**, 161–9.

Capps, R. W., Mateli, G. N. and Bradford, M. L. (1993) Reduce oil and grease content in wastewater, *Hydrocarbon Processing*, **72**, 5.

Churchley, J. (1998) Ozone for dye waste colour removal; four years operation at Leek STW, *Ozone Sci. Eng.*, **20**, 111–20.

Correia, V. M., Stephenson, T. and Judd, S. J. (1994) Characterisation of textile wastewaters — a review, *Environ. Technol.*, **15**, 917–29.

Costa, C. and Marquez, M. C. (1998) Kinetics of the PACT process, *Water Res.*, **32**, 107–14.

Dagnall, S. (1995) UK strategy for centralised anaerobic digestion, *Bioresource Technol.*, **52**, 275–80.

Garnham, G. W. (1997) The use of algae as metal biosorbents, in *Biosorbents for Metal Ions*, Wase, D. A. J. and Forster, C. F. (eds), Taylor and Francis, London, pp 11–37.

Gray, N. F. (1999) *Water Technology. An Introduction for Environmental Scientists and Engineers*, Arnold, London.

Hanaki, K., Matsuo, T. and Nagase, M. (1981) Mechanism of inhibition caused by long-chain fatty acids in anaerobic digestion process, *Biotechnol. Bioeng.*, **23**, 1591–610.

Harwood, H. J. and Geyer, R. P. (1964) *Biology Data Book*, Federation American Societies for Experimental Biology, Washington, DC, USA.

Hawkes, F. R., Hawkes, D. L. and Peck, M. W. (1985) Effect of mechanical separation on the anaerobic digestion of cattle slurry, *Proc. Conf. Advances in Fermentation*, Kings College, London. Supplement to *Process Biochem.*, September.

Hoefl, C., Sigl, G., Sprecht, O., Wurdack, I. and Wabner, D. (1997) Oxidative degradation of AOX and COD by different advanced oxidation processes: a comparative study with two samples of pharmaceutical wastewater, *Wat. Sci. Technol.*, **35(4)**, 257–64.

Hooda, P. S., Edwards, A. C., Anderson, H. A. and Miller, A. (2000) A review of water quality concerns in livestock farming areas, *Sci. Total Environ.*, **250**, 143–67.

Hrudey, S. E. (1984) The management of wastewater from the meat and poultry products industry, in *Surveys in Industrial Wastewater Treatment Vol. 1*, Barnes, D., Forster, C. F. and Hrudey, S. E. (eds), Pitman Publishing, London, UK, pp 128–208.

Hug, A., Harf, J., Vogel, F. and von Rohr, P. R. (1998) Deactivation of sewage sludge by wet oxidation (WO) using the LOPROX process, *Chem. Engng. Technol.*, **21**, 880–5.

Iggleden, G. J. and van Staa, R. (1984) Use of chemical dosing and dissolved air flotation in the recovery of protein from waste waters, *Inst. Chem. Eng. Symp. Series*, **84**, 381–96.

Lennox, S. D., Foy, R. H., Smith, R. V., Unsworth, E. F. and Smyth, D. R. (1998) A comparison of agricultural water pollution incidents in Northern Ireland with those in England and Wales, *Water Res.*, **32**, 649–56.

Lewin, D. C. and Forster, C. F. (1974) Protein recovery from dairy wastes by electroflotation, *Effl. Wat. Trt. J.*, 142–8.

MAFF (1998) *Codes of Good Agricultural Practices for the Protection of Water*, Ministry of Agriculture, Food and Fisheries, London.

Maloney, S. W., Boddu, V. M., Phull, K. K. and Hao, O. J. (1994) *TNT Red Water Treatment by Wet Air Oxidation*, USACERL Technical Report EP-95/01.

Mason, P. A. (1988) Dealing with silage effluent: technical opportunities, in *Silage Effluent*, Stark, B. A. and Wilkinson, J. M. (eds), Chalcombe Publications, Marlow, pp 21–36.

Matlock, M. M., Henke, K. R., Atwood, D. A. and Robertson, D. (2001) Aqueous leaching properties and environmental implications of cadmium, lead and zinc trimercaptotriazine (TMT) compounds, *Water Res.*, **35**, 3649–55.

Matlock, M. M., Henke, K. R. and Atwood, D. A. (2002) Effectiveness of commercial reagents for heavy metal removal from water with new insight for future chelate design, *J. Haz. Mat.*, **92**, 129–42.

McCay, G. and Allen, S. J. (1997) Low-cost adsorbents in continuous processes, in *Biosorbents for Metal Ions*, Wase, D. A. J. and Forster, C. F. (eds), Taylor and Francis, London, pp 183–220.

Miner, J. R., Humenik, F. J. and Overcash, M. R. (2000) *Managing Lifestock Wastes to Preserve Environmental Quality*, Iowa State University Press, Ames.

Misi, S. and Forster, C. F. (2002) Semi-continuous anaerobic co-digestion of agro-wastes, *Environ. Technol.*, **23**, 445–51.

Munter, R., Preis, S., Kallas, J., Trapido, M. and Veressinina, Y. (2001) Advanced oxidation processes (AOP): Water treatment technology for the twenty-first century, *Wat. Sci. Technol.*, **28(5)**, 354–62.

Narbaitz, R. M., Droste, R. L., Fernandes, L., Kennedy, K. J. and Ball, D. (1997) PACT^{TM} process for treatment of Kraft mill effluent, *Wat. Sci. Technol.*, **35(2–3)**, 283–90.

O'Brien, G. J. (1992) Estimation of the removal of organic priority pollutants by the powdered activated sludge treatment process, *Wat. Environ. Res.*, **64**, 877–83.

Perkowski, J., Kos, L. and Ledakowicz, S. (2000) Advanced oxidation of textile wastewaters, *Ozone Sci. Eng.*, **22**, 535–50.

Prado, J. and Esplugas, S. (1999) Comparison of different advanced oxidation processes involving ozone to eliminate atrazine, *Ozone Sci. Eng.*, **21**, 39–52.

Rajagopal, C. and Kapoor, J. C. (2001) Development of adsorptive removal process for treatment of explosives contaminated wastewater using activated carbon, *J. Haz. Materials*, **87**, 73–98.

Ratledge, C. (1992) Microbial oxidation of fatty alcohols and fatty acids, *J. Chem. Tech. Biotech.*, **55**, 399–400.

Reed, B. E., Robertson, J. and Jamil, M. (1995) Regeneration of granular activated carbon (GAC) columns used for removal of lead, *J. Environ. Eng.*, **121**, 653–62.

Rice, R. G. (1997) Applications of ozone for industrial wastewater treatment — a review, *Ozone Sci. Eng.*, **18**, 477–515.

Rivas, F. J., Beltran, F., Frades, J. and Buxeda, P. (2001) Oxidation of p-hydroxy benzoic acid by Fenton's reagent, *Water Res.*, **35**, 387–96.

Saunamäki, R. (1997) Activated sludge plants in Finland, *Wat. Sci. Technol.*, **35(2–3)**, 235–43.

Sayed, S., de Zeeuw, W. and Lettinga, G. (1984) Anaerobic treatment of slaughterhouse waste using a flocculant sludge UASB reactor, *Agric. Wastes*, **11**, 197–226.

Sheppard, A. J., Iverson, J. L. and Weihrauch, J. L. (1978) Composition of selected dietary fats, oils, margarines and butter, in *Handbook of Lipid Research, 1: Fatty Acids and Glygerides*, Kuksis, A. (ed), Plenum Press, New York, USA.

Sonntag, N. O. V. (1979) Reactions of fats and fatty acids, in *Bailey's Industrial Oil and Fat Products, Vol. 1, 4th Edition*, Swern, D. (ed), John Wiley and Sons, New York, USA.

Sparshott, M. B. (1991) Disposal of petrochemical plant effluents by Shell UK at Stanlow, *J. IWEM*, **5**, 599–607.

Thomas, A. E. and Paulicka, F. R. (1976) Solvent fractionated fats, *Chem. and Ind.*, Sept. 18, 774–9.

Virtue, W. A. and Clayton, J. W. (1997) Sheep dip chemicals and water pollution, *Sci. Total Environ.*, **194–195**, 207–17.

Walker, G. M. and Weatherley, L. R. (2000) Textile wastewater treatment using granular activated carbon adsorption in fixed beds, *Sep. Sci. Technol.*, **35**, 1329–41.

Wase, D. A. J. and Forster, C. F. (1997) (eds), *Biosorbents for Metal Ions*, Taylor and Francis, London.

Webb, A. R. and Hawkes, F. R. (1985) The anaerobic digestion of poultry manure: variation of gas yield with influent concentration and ammonium-nitrogen levels, *Agric. Wastes*, **14**, 135–56.

Webb, L. (1985) An investigation into the occurrence of sewage fungus in rivers containing paper mill effluents, *Water Res.*, **19**, 947–54.

Webb, L. J. (1994) Integrated pollution control of emissions from the pulp and paper industry, *Wat. Sci. Technol.*, **29(5–6)**, 123–30.

Wilczak, A. and Keinath, T. M. (1993) Kinetics of sorption and desorption of copper (II) and lead (II) on activated carbon, *Wat. Environ. Res.*, **65**, 238–44.

Wong, J., Maroney, P., Diepolder, P., Chiang, K. and Benedict, A. (1992) Petroleum effluent toxicity reduction — From pilot to full-scale plant, *Wat. Sci. Technol.*, **25(3)**, 221–8.

Ziminska, H. (1985) Protein recovery from fish wastewaters, *Proc. 5th Int. Symp. Agric. Wastes*, Chicago, Il.

11
The future

This chapter has been included not only to allow the author the indulgence of crystal ball gazing but also to ask, and in part answer, some of the questions which young water engineers will themselves have to ask as they become older engineers. Put another way, what will the sewage treatment works of the mid-21st century look like?

There would appear to be two questions of outstanding significance: the standards required for discharge and the future for biosolids (not sludge). Looking at the standards for discharge, the industry has moved from specifying BOD and suspended solids to standards which also require limits on COD, total nitrogen and total phosphorus. Thus, the basic wastewater treatment assets, which can be taken to mean the trickling filter and the activated sludge plant, are being required to achieve much more than that for which they were originally developed. Certainly, there are new or modified processes (Chapters 5 and 6) which may offer benefits in terms of meeting increasingly stringent demands, but one has to ask 'what comes next?', either from national legislation or from EU Directives. The crystal ball would suggest that trace organics might be next.

Trace organics

Endochrine disrupting chemicals
There has been much concern in the late 1990s about the relationship between trace organics in water and the occurrence of low sperm counts in both animal and human males, the incidence of testicular cancer and

Table 11.1. *Endocrine disrupting chemicals*

Category	Examples	Uses
Poly-chlorinated organics	Dioxins, PCBs	Byproducts, insecticides[a]
Organochlorine pesticides	Dieldrin, lindane	Insecticides[a]
Organotins	Tributyl tin	Anti-fouling agent
Alkyl phenols	Nonyl phenol	Surfactant precursor
Alkyl phenol ethoxylates	Nonyl phenol ethoxylate	Surfactant
Phthalates	Dibutyl phthalate	Plasticizer
Synthetic steroids	Ethinyl oestradiol	Contraceptive

the hermaphrodization of some animals and fishes (Montagnani et al., 1996; Tyler and Routledge, 1998). The endocrine system, in effect, controls the levels of hormones being secreted in the body. Oestrogens occur in both males and females albeit at different concentrations. In human males the circulatory concentration of oestradiol, the principal natural oestrogen, is less than 0·05 ng/litre, while for pre-menopausal females it is 0·06–0·2 ng/litre (Montagnani et al., 1996). Thus, the disruptions in the male reproductive systems were initially blamed on the use of the contraceptive pill and hormone replacement therapy. Certainly both natural and synthetic oestrogens are excreted from the human body, with increased levels occurring during pregnancy and hormone replacement therapy. It is also known that they are excreted in a conjugated form but are re-activated during passage through a biological treatment process. The activated sludge process will treat this type of compound. Typically, it would remove 70–88% of the oestradiol, 50–75% of the oestrone and 50–85% of the 17a-ethinyl-oestradiol. However, this would still leave a sufficient concentration of these hormones in the final effluent to affect the endocrine systems of sensitive fish.

In addition to the hormonal compounds, it is now clear that there are a range of chemicals which can mimic the oestrogens and cause disruption of the endocrine system (Table 11.1). Although some of these substances are now banned in the UK, others are in common use. For example, the alkyl phenol polyethoxylates are widely used as non-ionic surfactants both in domestic and industrial detergents, and it has been estimated that between 37 and 60% of all alkyl phenol polyethoxylates enter the aquatic environment. The effluents from sewage treatment works do contain endocrine disrupting chemicals and, therefore, they are being dispersed into the environment by this route. At the same time, it must be recognized that significant concentrations have been detected in river water upstream of sewage works' discharges. These are most likely to be associated with runoff from agricultural land. So, what can be done and what is the role of the water engineer? The Environmental Quality Standards (EQSs) have not been established for all the compounds known to be involved (Anon., 1998). This must be done so that treatment objectives can be derived. However, it is not always known how these chemicals are affected in conventional biological treatment plants. Certainly the alkyl phenol polyethoxylates undergo only partial degradation (Planas et al., 2002) and their degradation products, such as nonyl phenol, are also oestrogenic and less readily degradable (Ahel et al., 1994). Alkyl phenols are hydrophobic and, therefore, can accumulate in sludge. The fate of these compounds if these sludges are re-cycled to agricultural land is another problem. These are aspects of wastewater treatment which must be further examined. In addition, screening of other industrial chemicals must continue to ascertain whether there are other classes of chemical with these properties. Then there is the question of monitoring. Should it be in the environment itself or should it be at the sewage works' discharges? If it is the latter, the question of treatment needs to be addressed. It would be possible to remove trace organics from secondary treated sewage by using techniques such as membranes and granular activated carbon, but this would be expensive and would not deal with the entry of endocrine disrupting chemicals into the environment from other sources.

Toxicity
As well as the worries about endocrine disrupters, there is increasing concern about discharges containing chemicals which, either alone, or in combination with other compounds, could have adverse effects on elements of the food chains in surface waters. In addition, there is the additional philosophy which suggests that gross parameters such as BOD and COD have outlived their usefulness and that toxicity-based consents ought to be used (Wharfe and Tinsley, 1995). This is leading to the development of the concept of direct toxicity assessment (DTA), which would measure the overall effect on the receiving watercourse and would include antagonistic as well as synergistic effects of the mixtures (Hunt et al., 1992; Burgess et al., 2000). The basic concept of DTA is similar to that of Whole Effluent Toxicity (WET) which was introduced in the USA in 1991. Total toxicity cannot be derived theoretically even if all the individual toxicological effects are

Table 11.2. Toxicity testing techniques

Organism	Freshwater	Marine waters
Algae	*C. vulgaris*; 72/96 hour growth inhibition	*Phaeodactylum tricornutum*; 72/96 hour growth inhibition
Invertebrates	Water flea; 48 hour immobilization	Oyster larvae; 24 hour development inhibition
Fish	Rainbow trout; 96 hour mortality	Juvenile plaice or turbot; 96 hour mortality

known. A DTA, therefore, assesses the potential for harming the environment and puts a quantified limit on an effluent or chemical. As such, it has a role

- in water management;
- for establishing a consent protocol;
- for measuring the net toxic effect.

For rapid screening, much attention is being focused on the Microtox technique (Choi and Meier, 2001). Microtox is a commercially available test which uses a marine bacterium, *Photobacterium phosphoreum*. As it respires, it emits light. Toxic compounds reduce the respiration and, therefore, the amount of light emitted. Different dilutions of the chemical or wastewater are tested to give a plot of percentage reduction against concentration and from this a 50% reduction value can be obtained. However, the Microtox organism is not representative of the species which will be present in freshwater rivers. Toxicity tests which use representative species are available and it has been recommended that standard protocols based on *Chlorella vulgaris*, *Daphnia magna* and the rainbow trout, *Oncorhyncus mykiss*, to represent an alga, an invertebrate and a fish, should be used (Table 11.2) (Hunt et al., 1992). For routine assessments, the Microtox test could be used, but there would be a need to 'calibrate' it, both initially and periodically thereafter, against the most sensitive of the three main tests (Hunt et al., 1992).

Sewage biosolids

Although the processing and disposal of sewage sludge in the UK can be thought of as environmentally satisfactory under current policies, it is likely that more stringent environmental standards linked to a greater

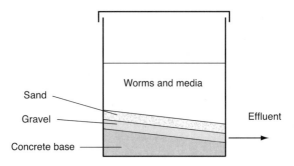

Fig. 11.1. Cross-section through a CLSV unit

emphasis on re-cycling will mean that the processes used for handling sludges will need to be re-thought. The fact that the land available for the re-cycling of sludge is declining must also be considered. Some potentially useful processes are already under investigation and it is these which will be discussed. Most of them seek to recover energy or some other useful product from the sludge.

One such route, which really is an extension of composting (Chapter 8), is the use of vermiculture. This is the use of earthworms to convert sludge into an odourless, useable product. This is not a new concept having first been examined in the late 1970s (Hartenstein, 1978), but it is only recently that it has been considered as a commercial possibility with five of the UK water companies supporting a pilot-scale evaluation of what is being called continuous liquid sludge vermi-stabilization (CLSV). Unlike earlier work, the current technology uses liquid sludge which eliminates the need and the cost of de-watering. The unit (Fig. 11.1) consists of a gravel and sand base for drainage with media above it to provide a habitat for the worms. Essentially, the media should provide a light aerobic structure which is compatible with the living requirements of the worms. Material which has been tested includes wood chip and plastic. The worm species being used is *Eisenia fotidia*. The generation time of this species is, typically, about 10–15 weeks. This means that there is an extended period, perhaps as long as a year, before such a unit could be considered to have reached a steady state. The process does produce a liquid effluent which would require to be treated in much the same way as the liquors from sludge de-watering.

The other potential options for processing sludge are, essentially, thermal treatments (Werther and Ogada, 1999). As has already been mentioned, sludge has a calorific value and it has been suggested that this could be recovered by co-combustion with fossil fuels at power stations (Barber, 2002). This is an approach which has already been adopted in Germany and the only apparent draw-back is that very rigorous gas cleaning, particularly flue gas de-sulphurization, would be required. However, this is something which is already installed at a number of UK power stations (Barber, 2002). Co-combustion with other materials has been discussed in the review by Werther and Ogada (1999).

Wet oxidation (Chapter 10) is a technology which achieves an oxidative degradation of organics without the need for costly gas cleaning equipment. Effectively, two grades of wet oxidation can be considered (Hudson and Lowe, 1996; Werther and Ogada, 1999)

- sub-critical which occurs at a pressure of 100 bar and a temperature of less than 374 °C;
- super-critical which occurs at a pressure of 218 bar and a temperature of more than 374 °C.

Sub-critical wet oxidation has the potential disadvantage that it produces a liquor with an appreciable COD and BOD which has to be treated. However, the conditions required for sub-critical operation can be achieved relatively easily and the review by Werther and Ogada (1999) lists some nine processes currently available for the treatment of sludges. Perhaps the best documented is the VerTech plant at Apeldoorn in the Netherlands which has the capacity to handle 22 800 tonnes dry solids/year (de Bekkar and van den Berg, 1993; Wanka, 1994; Werther and Ogada, 1999). This is a deep well reactor which is 950 mm in diameter, 1200 m deep and is cement lined (Fig. 11.2). The sludge circulates through two concentric tubes. The riser has a diameter of 340 mm and the downcomer has one of 200 mm. Oxygen is injected into the downcomer. The pressure at the bottom of the shaft is 100 bar and the temperature is controlled at 270/280 °C by the use of heat exchange fluid which surrounds the shaft. During start-up this is used to provide heat, but once the oxidation process is in progress it is used to ensure that the temperature does not become too high. The plant uses 45–55 tonnes oxygen/day and achieves a reduction in COD of 70%.

Super-critical wet oxidation, which should offer a greater degree of degradation, has been shown, at demonstration plant level, to give a 99·99% destruction of organic matter and will destroy persistent organics such as PCBs (Hudson and Lowe, 1996).

The generation of oil from sludge has been under investigation for some time (Bridle and Skrypski-Mantele, 1994). The conversion is

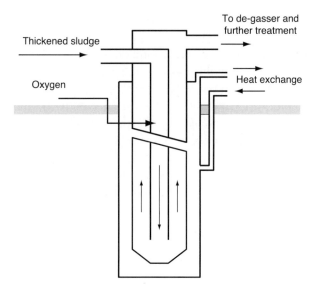

Fig. 11.2. Schematic diagram of the VerTech wet oxidation process

achieved by heating dried sludge at a temperature of 400–450 °C in the absence of oxygen. One such plant is operated in Australia. This Enersludge unit produces oil, a char and a combustible gas (Table 11.3), with the oil having properties of a medium fuel oil and the char having the characteristics of a low-grade coal. Taking a more generalized view of the oil from sludge technology, the yield of oil will range from 200 to 300 litres per dry tonne of sludge processes, with the maximum yield being achieved from raw sludge (Hudson, 1995).

The gasification of sludge is another option which can effect both solids' reduction and energy recovery (Furness et al., 2000). There are several commercial options available and these are discussed in detail in the review by Werther and Ogada (1999). Most of these are multi-stage processes in which the sludge is first pyrolysed at 400/450 °C in

Table 11.3. Expected products from the oil from sludge pyrolysis of raw sludge

Product	Yield (%)	Calorific value (MJ/kg)
Oil	30	38
Char	50	12
Gas	10	10
Water	10	4

The future

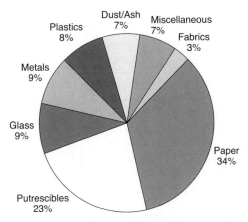

Fig. 11.3. Typical composition of municipal solid waste

the absence of oxygen and the products of this stage are then heated to 800–1400 °C. The gasification stage can have an input of air or it may be purely a thermal process. The latter gives a product gas with the highest calorific value as it is not diluted. These stages can be carried out in separate reactors or in a single unit with a temperature gradient across the reactor which allows for drying, pyrolysis and gasification. The gasifier units can be based on fixed bed, fluidized bed or moving bed reactors. The product is a fuel gas containing hydrogen and hydrocarbons which can have a calorific value as great as 29 MJ/m^3. In mainland Europe various designs of pyrolysis/gasification systems are being operated on a large scale, processing a variety of waste solids including sewage sludge (Werner and Ogada, 1999; Jaeger and Mayer, 2000). In the UK, Anglian Water are currently examining the performance of an integrated downdraft dryer/gasification system at their Broadholme Sewage Treatment Works. The Syngas produced by this process has a calorific value of more than 3·5 MJ/m^3 (Riches and Evans, 2002). A plant has also been installed at the Nash treatment works in Wales with a nominal feed rate of 500 kg/h of dried sewage sludge and screenings.

Any discussion of processes which can re-cycle energy and which have the potential for handling sludge in the future must also consider municipal solid waste (MSW). This has an appreciable organic content (Fig. 11.3) which could well be used to supplement that present in sewage sludge. Thames Waste Management, a part of Thames Water, are already doing this. The MSW is processed to remove non-degradable material such as plastic, metal and glass, and the residue is then slurried and digested anaerobically with sewage sludge using the mesophilic digesters which are part of Thames Water's assets.

Table 11.4. Estimated quantities of biosolids re-cycled to agricultural land in the UK (Hickman, 1999)

Biosolids	Mass to land (million tonnes/year)	
	Fresh weight	Dry solids
Sewage sludge		0·5
Cattle manure	73·3	12·0
Pig manure	10·4	1·0
Poultry manure	4·4	2·1
Sheep manure	2·6	0·6

The question of animal manures will also need consideration. Much of this is re-cycled to agricultural land (Table 11.4), but will there be demands in the future for microbiological standards similar to those applied to sewage sludge? If this were to happen, it is unlikely that the necessary processing could, or even should, be done on the farm. Rather, it should be done at some centralized facility. This is not a new concept (Forster and Jones, 1976; Dagnall, 1995), but it is not one which has been put into practice in the UK. However, the recognition that sustainability is an important aspect of waste management has led to the gradual development of the specific waste re-cycling centres which could handle a range of organic wastes. For example, a new plant has recently been commissioned in Devon based on a catchment with a 15 km radius. It aims to process 146 000 tonnes/annum of cattle, pig and poultry manures plus organic food waste. After mixing, the wastes will be pasteurized (one hour at 70 °C) before being digested at 37 °C in one of two 4000 m^3 digesters (Fig. 11.4). The digestion time will be 20 days

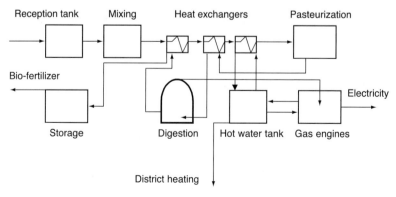

Fig. 11.4. Schematic diagram of the Holsworthy plant in Devon

Table 11.5. Summary of the Kristianstad biogas plant, 1998

Inputs	Tonnes
Pig manure	20 800
Cattle manure	20 000
Poultry manure	400
Household waste	3 100
Abattoir waste	24 600
Distillery waste	900
Vegetable waste	1 400
Total	71 200

Outputs	Tonnes
Digested sludge for fertilizer	67 150
Biogas	4 000
Rejects	50
Total	71 200

and the 'budgeted' gas production is 3.9×10^6 m^3 methane/year (Fink, 2002).

A similar approach is used at Kristianstad in Sweden where the biogas plant handles waste from industry and agriculture, as well as organic household waste. The solid wastes are pasteurized by holding them at 70 °C for one hour before digestion, which takes place at 38 °C with a retention time of 20–24 days. The performance of the plant is summarized in Tables 11.5 and 11.6. Particular features about this plant include the facts that

- household waste is sorted in the home but there still needs to be a facility for removing metals;

Table 11.6. Energy balance for the Kristianstad biogas plant, 1998

	MW h
Gross biogas recovery	20 000
Process heating	2 100
Sales to district heating	17 900
Electricity bought in	540

- industrial wastes are charged at a rate less than that charged at a landfill site;
- farmers who deliver manure receive fertilizer free of charge.

Activated sludge process

As has been discussed in Chapter 4, the activated sludge process can suffer from operational problems, bulking and foaming, which disrupt its efficiency. Essentially, these stem from an incomplete understanding of the microbiology of the process. There is a reasonable understanding about the way in which the nitrogen oxidizers operate. The effect of temperature is well defined. The interactions with alkalinity are known, as are the values of the kinetic constants. Knowledge about the carbon oxidizers, on the other hand, is very limited. There are really no data about whether the various species interact with one another, and, if so, in what way. Kinetic constants are available for some species, but all too often they are quoted for a generic species 'heterotrophic bacteria' or for specific species grown on substrates such as glucose or acetate, which are not very representative of sewage. In other words, the 'work horses' of the activated sludge process are too poorly defined for process control purposes.

The situation is gradually changing and sludge microbiology is receiving more attention (Seviour and Blackall, 1999), but, mostly, this is focused on the filamentous species. However, there is yet to be a clear differentiation between a filament dominated sludge which produces stable foam and one which does not settle well. It is tempting to wonder whether there are some parameters other than filament length which need to be considered. Activated sludge is negatively charged and this charge is usually associated with the extracellular polymeric substances (EPSs) which make up the sludge matrix (Forster, 1971). Its composition will depend on factors such as the dissolved oxygen concentration (Palmgren et al., 1998), the sludge age (Liao et al., 2001) and the composition of the feed (Boyette et al., 2001). Essentially, an EPS is a mixture of carbohydrates, proteins and nucleic acids, and the negatively charged functional groups in these compounds generate the charge. These can be the carboxyl groups of proteins (Dignac et al., 1998) or acidic sugars (Steiner et al., 1976). Surface charge has been associated with the settlement of activated sludge by previous workers (Forster, 1971; Lee, 1997). However, the more usual opinion is that it is the total filament length which controls settlement. Filaments are an essential element in the activated sludge floc structure in that they act as a 'backbone' for the floc (Sezgin et al., 1978). It is when they grow out from the floc that problems occur. However, no studies have been made to relate both filaments and surface charge to settlement. The genetic characteristics of the filamentous species is gradually being documented (Seviour and Blackall, 1999). What is needed is an understanding of their physico-chemical characteristics as well.

Recently, work has been reported describing the use of minerals, such as clay, to counteract bulking and foam formation by modifying the sludge surfaces (Stratton et al., 1998; Eikelboom and Grovenstein, 1998). The argument that sludge surfaces are related to the overall characteristics of the sludge is not new and interactions which could explain the phenomena of bulking and foaming have been reported (Forster, 1996) (Fig. 4.33). The rationale for using clay-type minerals is that they would impart a degree of hydrophilicity to the sludges and, thus, reduce their foaming potential, and would add weight to enhance their settlement properties. The hypothesis is quite sound as foam-forming sludges are hydrophobic (Khan et al., 1991). A number of potential antifoaming reagents were tested at a final concentration of 500 µg/ml using Rhodococcus rhodochrous, which is a foam-producing nocardioform (Stratton et al., 1998). The results are summarized in Fig. 11.5 and show that talc, bentonite and zeolite were the most effective. However, these procedures have not been optimized. If they could be and if a sound understanding of what controlled the hydrophobicity/hydrophilicity balance of sludge surfaces could be developed, then it is possible that the problems of bulking and foaming might cease to plague plant operators.

Substrate balancing by the addition of nitrogen and phosphorus is a well established practice, but the use of trace elements or organic compounds is found only occasionally. It was established in 1972 that a range of simple organic compounds such as glucose at concentrations of 5–10 mg/litre could significantly reduce the hydraulic retention time required to achieve the treatment of coke oven liquors (Catchpole and Cooper, 1972). However, the biochemical reasons for the success of these 'growth factors' was never established and their use did not receive widespread attention. More recently, there has been considerable interest in the use of micro-nutrients (Clark and Stephenson, 1998; Burgess et al., 1999). The use of organic micro-nutrients certainly appears to have the potential for enhancing the performance of activated sludge plants. For example, in one study, the COD removal rates were increased from 1·34 kg COD/kg MLSS per day to 4·24 kg COD/kg MLSS per day by the

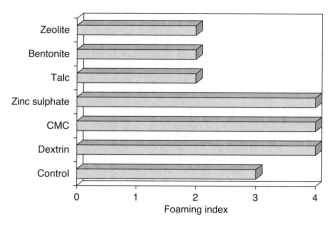

Fig. 11.5. The effect of added chemicals on the foaming ability of R. rhodochrous (Stratton et al., 1998)

addition of 0·01 µg/litre pyridoxine (Burgess et al., 1999). This type of work must be thought of as one of the advances of the last decade, but it is a technique which is still in its infancy.

The use of chemical supplements can also be used to reduce the sludge yield by uncoupling the catabolic reactions, which degrade molecules and produce free energy, from the anabolic reactions which use the energy and build-up cell material (Mayhew and Stephenson, 1997). Again this is an area which still requires a considerable amount of work, but, when it is considered that up to 65% of wastewater treatment operating costs are associated with the handling and disposal of sewage sludge (Bruce et al., 1990), it is work which should be encouraged. Certainly, metabolic uncoupling, whether by chemical addition or by process manipulation, could be considered as an advance for the next decade.

It could be argued that, by 2050, the activated sludge plant will have become superceded by processes such as some of those described in Chapter 5. Membrane bioreactors or biological aerated filters would not suffer from foaming or bulking and, therefore, the developments which have been suggested might be irrelevant. However, the capital investment, certainly in large treatment plants, probably means that the activated sludge process will survive for some time, although in what form it is difficult to say. Improved knowledge of the biological and physico-chemical characteristics of the sludge and its component micro-organisms would mean that mathematical models would become more accurate and enable plants to be designed on a less empirical basis. It could also mean that the sludge could be constructed with a specialized and specific ecology, possibly including genetically engineered bacteria (Nusslein et al., 1992).

References

Ahel, M., Giger, W. and Koch, M. (1994) Behaviour of alkylphenol polyethoxylate surfactants in the aquatic environment. 1. Occurrence and transformation in sewage treatment, *Water Res.*, **28**, 1131–42.

Anon. (1998) *Endocrine-disrupting Substances in the Environment: What Should be Done?* Environment Agency, Bristol.

Barber, W. P. (2002) Potential for sewage sludge co-combustion in the UK based on German experience, *J. CIWEM*, **16**, 270–6.

Boyette, S. M., Lovett, J. M., Gaboda, W. G. and Soares, J. A. (2001) Cell surface and exopolymer characterization of laboratory stabilized activated sludge from a beverage bottling plant, *Wat. Sci. Technol.*, **43(6)**, 175–84.

Bridle, T. R. and Skrypski-Mantele, S. (1994) The Enersludge™ process, in *Proc. Symp. Innovative Technologies for Sludge Utilisation and Disposal*, CIWEM, December.

Bruce, A. M., Pike, E. B. and Fisher, W. J. (1990) A review of treatment process options to meet the EC sludge directive, *J. IWEM*, **4**, 1–13.

Burgess, J. E., Quarmby, J. and Stephenson, T. (1999) Micronutrient supplements to enhance the biological wastewater treatment of phosphorus-limited industrial effluents, *Trans. I. Chem. E.*, **77 Part B**, 199–204.

Burgess, J. E., Longhurst, P. J., Quarmby, J. and Stephenson, T. (2000) Innovational adaptation in the UK water industry: a case study introducing DTA, *Technovation*, **20**, 37–45.

Catchpole, J. R. and Cooper, J. L. (1972) The biological treatment of carbonization effluents — III New advances in biochemical oxidation of liquid wastes, *Water Res.*, **6**, 1459–74.

Choi, K. and Meier, P. G. (2001) Toxicity evaluation of metal plating wastewater employing the Microtox® assay: A comparison with cladocerans and fish, *Environ. Toxicol.*, **16**, 136–41.

Clark, T. and Stephenson, T. (1998) Effects of chemical addition on aerobic biological treatment of municipal wastewaters, *Environ. Technol.*, **19**, 579–90.

Dagnall, S. (1995) UK strategy for centralised anaerobic digestion, *Bioresource Technol.*, **52**, 275–80.

de Bekkar, P. H. A. M. J. and van den Berg, J. J. (1993) Modelling and construction of a deep well aqueous phase oxidation process, *Wat. Sci. Technol.*, **27(5/6)**, 457–68.

Dignac, M. F., Urbain, V., Rybacki, D., Bruchet, A., Snidaro, D. and Scribe, P. (1998) Chemical description of extracellular polymers: implication on activated sludge floc structure, *Wat. Sci. Technol.*, **38(8–9)**, 45–53.

Eikelboom, D. H. and Grovenstein, J. (1998) Control of bulking in a full scale plant by the addition of talc (PE 8418). *Wat. Sci. Technol.*, **37**, 297–302.

Fink, J. (2002) Renewable energy, *Landwards*, **57(5)**, 10–12.

Forster, C. F. (1971) Activated sludge surfaces in relation to the sludge volume index, *Water Res.*, **5**, 861–70.

Forster, C. F. (1996) Aspects of the behaviour of filamentous microbes in activated sludge. *J. CIWEM*, **10**, 290–4.

Forster, C. F. and Jones, J. C. (1976) The Bioplex concept, in *Food from Waste*, Birch, G. G., Parker, K. J. and Worgan, J. T. (eds), Applied Science Publishers, London, pp 278–89.

Furness, D. T., Hoggett, L. A. and Judd, S. J. (2000) Thermochemical treatment of sewage sludge, *J. CIWEM*, **14**, 57–65.

Hartenstein, R., (1978) Physicochemical changes effected in activated sludge by the earthworm *Eisenia foetida*, *J. Environ. Qual.*, **10**, 372–6.

Hickman, G. A. W. (1999) Sustainable beneficial biosolids recycling to land — Future issues, in *Sludge 9, A One-day Seminar on Developments in Sludge Treatment and Disposal*, University of Surrey, September.

Hudson, J. (1995) Treatment and disposal of sewage sludge in the mid-1990s, *J. CIWEM*, **9**, 93–100.

Hudson, J. A. and Lowe, P. (1996) Current technologies for sludge treatment and disposal, *J. CIWEM*, **10**, 436–41.

Hunt, D. T. E., Johnson, I. and Milne, R. (1992) The control and monitoring of discharges by biological techniques, *J. IWEM*, **6**, 269–77.

Jaeger, M. and Mayer, M. (2000) The Noell Conversion Process — a gasification process for the pollutant-free disposal of sewage sludge and the recovery of energy and materials, *Wat. Sci. Technol.*, **41(8)**, 37–44.

Khan, A. R., Kocianova, E. and Forster, C. F. (1991) Activated sludge characteristics in relation to stable foam formation, *J. Chem. Tech. Biotech.*, **52**, 383–92.

Lee, B. C. (1997) *The influence of nutrients on floc physicochemical properties and structure in activated sludge process*, M.A.Sc. Thesis, University of Toronto.

Liao, B. Q., Allen, D. G., Droppo, I. G., Leppard, G. G. and Liss, S. N. (2001) Surface properties of sludge and their role in bioflocculation and settleability, *Water Res.*, **35**, 339–50.

Mayhew, M. and Stephenson, T. (1997) Low biomass yields activated sludge: a review, *Environ. Technol.*, **18**, 883–92.

Montagnani, D. B., Puddefoot, J., Davie, T. J. A. and Vinson, G. P. (1996) Environmentally persistent oestrogen-like substances in UK river systems, *J. CIWEM*, **19**, 399–406.

Nusslein, K., Maris, M., Timmis, K. and Dwyer, D. (1992) Expression and transfer of engineered catabolic pathways harboured by *Pseudomonas* sp. introduced into activated sludge microorganisms, *Appl. Environ. Microbiol.*, **42**, 3380–6.

Palmgren, R., Jorand, F., Nielsen, P. H. and Block, J. C. (1998) Influence of oxyygen limitation on the cell surface properties of bacteria from activated sludge, *Wat. Sci. Technol.*, **37(4–5)**, 349–52.

Planas, C., Guadayol, J. M., Droguet, M., Escalas, A., Rivera, J. and Caixach, J. (2002) Degradation of polyethoxylated nonyl phenols in a sewage treatment plant. Quantitative analysis by isotopic dilution-HRGM/MS, *Water Res.*, **36**, 982–8.

Riches, S. and Evans, I. (2002) Biosolids gasification studies, *Sewage Sludge Treatment and Use*, IQPC, London, September.

Seviour, R. J. and Blackall, L. L. (1999) (eds) *The Microbiology of Activated Sludge*, Kluwer Academic Publishers, Dordrecht, The Netherlands.

Sezgin, M., Jenkins, D. and Parker, D. S. (1978) A unified theory of filamentous activated sludge bulking. *J. Wat. Pollut. Control Fed.*, **50**, 362–81.

Steiner, A. E., McLaren, D. A. and Forster, C. F. (1976) The nature of activated sludge flocs, *Water Res.*, **10**, 25–30.

Stratton, H., Seviour, B. and Brooks, P. (1998) Activated sludge foaming: what causes hydrophobicity and can it be manipulated to control foaming? *Wat. Sci. Technol.*, **37**, 503–10.

Tyler, C. R. and Routledge, E. J. (1998) Oestrogenic effects in fish in English rivers with evidence of their causation, *Pure Appl. Chem.*, **70**, 1795–804.

Wanka, H. (1994) VerTech — Wet air oxidation of sewage sludge, *Fett Wissenschaft Technologie*, **96**, 527–9.

Werther, J. and Ogada, T. (1999) Sewage sludge combustion, *Prog. Energy Combust. Sci.*, **25**, 55–116.

Wharfe, J. R. and Tinsley, D. (1995) The toxicity-based consent and the wider implication of direct toxicity assessment to protect aquatic life, *J. CIWEM*, **9**, 526–30.

Index

Note: Page numbers in *italics* refer to illustrations, tables and diagrams.

acetogenic bacteria 218
acidogenic bacteria 216–217, 218
   in anaerobic digestion 254
   kinetic constants 254
activated carbon
   see also granular activated carbon;
      powdered activated carbon
   industrial wastewaters treated with
      282–283, 286–289
activated sludge process 67–108
   see also oxidation ditches
   aeration devices 73–77
      fine-bubble diffused-air system 73,
         74, 83–84
      jet aeration system 73, 76
      mechanical surface aerators 73,
         74–75
      submerged turbine aerator 73, 76
   aeration plant design 73–85
   aeration tank configuration 86, 90–92
   aerator efficiencies 77–79
   anoxic zone at inlet 86, 87, 164–167
      hydraulic retention time for 87,
         165, 166
   bacteria in 17, 68–69
   biomass in 67–68
   bulking in 68, 99
      control of 99–103
   characteristics 44
   compared with biological aerated
      filters 314
   compared with membrane bioreactors
      146–147, *147*, 314
   compared with trickling filters 44
   complete-mixing regime in aeration
      tank 90, *91*, 92
   de-nitrification performance 165
   design options 107–108
   dissolved oxygen control in 75, 82–83
   effect of iron salts 172–173
   essential nutrients requirements
      86–87
   factors affecting amount of sludge
      produced 93–95
   final settlement tanks 87–90
   fine-bubble diffused-air (FBDA)
      system 73, *74*, 83–84
      aeration efficiency 73, 75, 77, 78
      anoxic zone at inlet end 86, 87
      degree of mixing estimated for 92
      dissolved-oxygen control 82–83
      in hybrid aeration system 83–84, 85
      optimization of aeration efficiency
         80–82
   foam formation in 68, 103–104
      control of 104–105
   future developments 312–315
   hybrid aeration systems 83–85

activated sludge process (*continued*)
  hydraulic retention time 20, 72–73
  mechanical surface aerators 73, 74–75, 83
    aeration efficiency 78
    dissolved oxygen control 83
    in hybrid aeration systems 84
  microbiology 67–69, 312–313
  mixing regimes in aeration tank 90–92
  nitrification in 161–162
  operational problems 95–105
    bulking 68, 99–103
    foam formation 68, 103–105
  optimization of aeration efficiency 80–82
  organic loading rate 70–71
  oxygen requirements 79–80
  oxygen transferred by aerators 79
  oxygenation capacity 77
  performance curves 72, 162
  phosphate removal from effluent 170–178
    Bardenpho process 174–175
    by biological process 173–178
    by chemical treatment 170–173
    by post-precipitation 172
    by pre-precipitation 171
    by simultaneous precipitation 171
  plug-flow configuration of aeration tanks 86, 90, 91, 92
  powdered activated carbon used in 282–283
  process variables 70–73
  selectors used to control bulking and foaming 101–102, 105
    aerobic selectors 101–102
    anoxic selectors 102
  sludge age effects 71–72
  sludge recycle rate 94–95
  sludge settleability 85–87
  tapered aeration system 80–81
  variation of sludge wastage rate 94
ADAS Safe Sludge Matrix 208, 237
adenosine triphosphate (ATP) 15
adsorption isotherms 288
adsorption processes
  industrial wastewaters treated by 286–289

  oxygen-rich air generated by 135–136
advanced oxidation 289–290
advanced treatment of sludge 208–209
aerated channels (for grit removal) 30, 31, 32
aeration, purpose in activated sludge process 73
aeration devices (in activated sludge process) 73–77
  types 73, 74–76
aerobic digestion, enhanced treatment of sludge by 228–230
agricultural use of sewage sludge 237–241
  application rate calculations 238–239
  and endocrine-disrupting chemicals 304
  legislation covering 206–207, 237
  in various European countries 204
agricultural wastes 294–297
algae, in trickling filters 46
alkanes, oxidation of 280
aluminium salts
  industrial wastewaters treated with 284, 285
  phosphates removed using 169, 172
American Petroleum Institute (API) separator 281
ammoniacal-nitrogen 1
  amount from one human 2
  in industrial wastewaters 284
  leaching into surface waters 240
  oxidation of
    in biological aerated filters 121, *122*
    in membrane bioreactors *147*
    in oxidation ditches 141
    in reed beds 144, *195*
    in trickling filters 45–46, *52*, 55
  reactions with chlorine 197
  in sewage sludge 237
amoebae 21–22
anaerobic baffled reactor 264–265
anaerobic digestion of liquid wastewaters 253–270
  advantages 265, 266
  basic considerations 266–268
  compared with aerobic processes 265–266

Index

costs 265
disadvantages 266
microbiology 253–255
modes of operation 268–269
prevalence 269–270
reactor design 255–265
  anaerobic filters 257–258
  contact digesters 255–257
  upflow sludge blanket reactors 258–261
anaerobic digestion of sludge
  acetogenic stage *217*, 218
  acidogenic stage 216–218, *217*
  dual digestion 223–225
  enhancement of 225–228
    by enzymatic hydrolysis 227–228
    by thermophilic digestion 227
    by ultrasound 226–227
  factors affecting 220, 221–222
  heat input required 219
  hydrolysis stage 216, *217*, 225
  materials causing inhibition 221–222
  mesophilic digestion 216–223, 225
  methanogenic stage *217*, 218
  microbiology 216–218
  sludges stabilized by 216–228
  thermophilic digestion 227
  two-stage systems 223–225
anaerobic digestion of slurried animal wastes 297
anaerobic expanded bed digester 263–264
anaerobic filters 257–258
  downflow filter 257, 258
  upflow filter 257, 258
anaerobic fluidized bed digester 263–264
  performance 264
anaerobic sludge digesters
  design and operation of 218–223
  egg-shaped 223, *224*
  fixed-roof tanks 219, *221*
  floating-roof tanks 219, *221*
  performance data 222
animal manures
  processing of 310–311
  re-cycling of *310*
anoxic zone (in activated sludge process) 86, 87

de-nitrification in 164–167
anti-foams 105
autothermal thermophilic aerobic digestion (ATAD) process 228–230
  handling characteristics of sludges 229–230
  temperature cycles in *229*

B2A (biological aerated filter) system 122
  performance *122*
bacteria 16–21
  see also filamentous microbes
  in activated sludge process *17*, 68–69, 95–98
  in anaerobic digestion 216–217, 218
  dimensions *18*
  and disinfection 195–202
  growth 18–20
  removal by membrane bioreactors 146
  in sewage sludge 206
  in trickling filters 45–46
bacteriophages 23
Bardenpho process 174–175
Bathing Water Directive 42, 195
  microbiological standards *196*
  UK compliance *196*
belt thickeners (for sludge thickening) 211
best available techniques not entailing excessive cost (BATNEEC) 274, 275
biochemical oxygen demand (BOD) 1
  in agricultural wastewaters 294
  from one human 1
  in industrial wastewaters 2, 276, 277, 284
  in municipal sewage 2, 276
  oxidation of
    in biological aerated filters 121
    in biological contactors *116*
    on grass plots 185
    in lagoons 185
    in membrane bioreactors 147
    in oxidation ditches 141
    in primary settlement tanks 34
    in reed beds 144, 195
    in trickling filters 45, 52, 55, 63–64

321

biocides, to suppress foaming 105
biodegradability, of fats and oils 280
biofilters, for odour control 153
biogas
   calorific value 219
   formation of 218, 224, 256, 258, 259, 261, 262, 263, 265
Biogran Natural (dried sludge) 248
   composition 248
biological aerated filters (BAFs) 118–122
   compared with activated sludge process 314
   de-nitrification in un-aerated version 168
   in enhanced biological treatment 283
   industrial wastewaters treated using 121–122
   loading rates for 120
   media used 118, 119
   performance 120–121
biological contactors 113–116
   rotating 113, 114–115, 116
   submerged 113, 115–116, 117
biological treatment
   of industrial wastewaters 275–281
   phosphate removal using 173–178
biology 14–23
bio-oxidation
   bacteria in 20, 45–46, 68–69, 95–98
   for odour control 148–154
   processes for 14, 113–154
      see also activated sludge process; biological contactors; deep shaft process; membrane bioreactors; oxidation ditches; oxygen-aerated systems; reed beds; sequencing batch reactors; submerged filters; trickling filters
bioscrubbers, for odour control 153
biosolids 305–312
   amount in UK 310
biostandards, for sludge 207
brewery wastewater
   composition 276
   treatment by anaerobic digestion 257, 261, 262, 269

bulking in activated sludge process 68, 99
   control of 99–103

calcium phosphate
   formation during phosphate removal 178–179
   uses 178
calcium salts, industrial wastewaters treated with 284, 285
capillary suction time (CST)
   measurement of 233–235
   typical values 229, 235
Carrousel system (oxidation ditch) 138, 139
CASS lift-clear decanters (in SBR plants) 128, 129
catalytic oxidation, odour compounds 153–154
centrifugation, sludge thickening by 210, 211
charging policy for treatment 8–10
chemical oxidation
   industrial wastewater treated by 286, 289–291
   odour control by 154
chemical oxygen demand (COD) 1
   hard COD 282
   in industrial wastewaters 2, 276, 284
   limit under UWWT Directive 5, 282
   in municipal sewage 2, 276
   oxidation of
      in anaerobic digestion reactors 256, 257, 258, 261, 263, 264, 265, 268
      in biological aerated filter 122
      in deep shaft process 126
      by ozone 283
      by wet air oxidation 290–291
   ratio to BOD 2
chemical treatment, phosphate removal using 170–173
chloramines, formation of 197
chlorinated compounds 198
chlorination, to suppress foaming 105
chlorine
   disinfection by 196–198
   reactions with ammoniacal-nitrogen 197

Index

reactions with water 197
ciliates (protozoa) 22, 46, 69
circular primary settlement tanks 36–38
circular trickling filters 49–50
coagulants, industrial wastewaters
    treated by 284–285
coastal waters, discharges to 41–42
co-combustion 307
Combi sludge drying system 247–248
composting of sludge 245–247
  see also vermiculture
  in aerated static piles 246
  agitated bay system 246
  in windrows 246
consents 3–10
  for trade effluents 6–7, 281
constant-velocity grit channels 29–30
constructed wetlands 142–145
  see also reed beds
contact digesters (for anaerobic digestion
    of wastewaters) 255–257
  performance data 256
continuous liquid sludge
    vermistabilization (CLSV) 306
corrugated plate interceptor 281, 282
Cryptosporidium oocysts
  removal of 56
  in sewage sludge 206

Dangerous Substances Directive 5
  heavy metals listed 6
deep bed filters 192
  compared with rapid gravity filters
    192
  de-nitrification in 168
deep shaft process 122–126
  COD removal by *126*
  design criteria 124–125
  start-up 123, *124*
de-nitrification 164–168
  anoxic zone for 164–167
  factors affecting rates 165–167
  fixed film de-nitrifying unit for 168
de-odorants 154
detritors 30, *31*
de-watering of sludge 230–237
  see also sludge de-watering
DHV Crystalactor^R process 178–179, *179*

quality of product *179*
diauxic growth 20, *21*
diluted sludge volume index (DSVI) 85
  and design of activated sludge final
    settlement tanks 87–88
direct toxicity assessment (DTA)
    304–305
disinfection 195–202
  aims 196
  by chlorine 196–198
  by ozone 198–200
  by ultraviolet radiation 200–202
dispersion number (in activated sludge
    aeration tank) 91, 92
  and sludge settlement characteristics
    101
disposal of residues 297
disposal of sludge 237–247
  see also sludge disposal
dissolved air flotation (DAF)
  fat/oil removal using 281
  industrial wastewaters treated using
    282
  sludge thickening by 212–213
dissolved oxygen (in activated sludge
    process)
  control of 75, 82–83
  minimum requirements 86
domestic water use, daily amount per
    person 2
downflow anaerobic filter 257
  performance *258*
drinking water standards 4
dry scrubbing/absorption, odour control
    by 154
drying of sludge 247–248
Dynasand filter 193–194

earthworms *see* vermiculture
ecological trophic pyramid, for biofilm of
    trickling filter *45*
eco-stratification 45, 161
Eikelboom technique for identification of
    filamentous microbes 95, 96, 97
electrostatic precipitation 243
endocrine-disrupting chemicals 302–304
  effects 302–303
  examples *303*

enhanced biological treatment, of
  industrial wastewaters 282–283
enhanced oxidation, of industrial
  wastewaters 289–291
enhanced treated sludge, biostandards
  for 208
enhanced treatment of sludge 228–230
  by aerobic digestion 228–230
  by lime treatment 230
Environment Agency
  consents 3
  eutrophic conditions defined by 160
  incinerator stack emission limits 243
  as regulator of waste management
    practice 206
  River Ecosystem classes 10
Environmental Act (1995) 6, 206
Environmental Protection Act (1990)
  57, 149, 206, 274
environmental quality standards (EQSs) 6
enzymatic hydrolysis, anaerobic digestion
  enhanced by 227–228
enzymes 15
Escherichia coli
  dimensions 18
  inactivation by ozone 199, 200
  inactivation by UV disinfection 202
  in sewage sludge 206
European Directives 3, 4–5, 41, 42
  on agricultural use of sludge 206, 240
  Bathing Water Directive 42, 195, 196
  on dangerous substances 5, 6
  on hazardous wastes 245
  on landfill 245
  Urban Wastewater Treatment
    (UWWT) Directive 3, 4–5, 41,
    160, 161, 176, 204–205
eutrophic conditions, factors affecting
  160
expanded bed anaerobic digester
  263–264
expanded granular sludge bed (EGSB)
  reactor 261–263
  examples of various plants 262
  performance of typical plants 263
extracellular polymeric substances
  (EPSs) in activated sludge process
  68, 312

faecal coliforms
  see also *Escherichia coli*
  inactivation by UV disinfection 201,
    202
  limit for bathing water 196
  limit for potable water 4
farm animal slurries, BOD data 294
fat-traps 281
fats, oils and greases (FOG) 278
  fatty acid composition 278, 278
  foaming promoted by 104
  in industrial wastewater 277–278
  non-degradable proportion 280–281
  removal of
    during grit removal 30, 31, 32
    in primary settlement tanks 38
fatty acids 278
  examples 279
  in fats and oils 279
  β-oxidation of 280
Fenton's reagent 289
filamentous microbes in activated sludge
  17, 68–69, 95–98, 312–313
  abundance scoring 97
  environmental requirements 98, 99
  identification of 95, 96, 97
  operational problems caused by
    68–69, 95, 99–105
  ranking in bulking ability 99
fill-and-draw reactors 126–128, 137
  see also oxidation ditches; sequencing
    batch reactors
filter press
  de-watering of sludge by 231–235
  fabrics used 231–233
filterability of sludges 233
  methods of measuring 233–235
  typical values 233, 235
final settlement tanks
  for activated sludge process 87–90
  for trickling filters 55
fine-bubble diffused-air (FBDA) system
  (in activated sludge process) 73, 74,
  83–84
  aeration efficiency 73, 75, 77, 78
  degree of mixing estimated for 92
  dissolved-oxygen control 82–83
  in hybrid aeration system 83–84, 85

optimization of aeration efficiency 80–82
fish farms, pollution by 297
fish kills, effect of agricultural wastes 294–295, *295*, *296*
flagellates (protozoa) 22
flies on trickling filters 57–58
　control of 57, 58
flotation methods
　fat/oil removal by 281
　sludge thickening by 212–213
flow to full treatment (FTFT) 32
　separation of flows in excess of 33–34
fluidized bed anaerobic digester 263–264
　performance *264*
fluidized bed incinerators 241, *242*
foaming in activated sludge process 68, 103–104
　control of 104–105, 313, *314*
food:mass (F:M) ratio
　see also organic loading rate
　for activated sludge process 70
France, sludge disposal options *204*
Freundlich adsorption isotherm 288
Fundamental Intermittent Standards (FISs) 10–11
fungi, in trickling filters 46
future developments 302–315

gasification of sludge 308–309
Germany, sludge disposal options *204*
Giardia spp.
　removal by trickling filters 56
　in sewage sludge 206
glucose, degradation of 15
granular activated carbon (GAC),
　industrial wastewaters treated with 286–289
grass plots 184–186
　compared with reed beds 186
　performance data *185*
　solids removal by 184–186
gravity thickening of sludge 209–210
Greece, sludge disposal options *204*
grit removal 29–31
　amount collected 31
　by aerated channels 30, *31*, *32*
　by constant-velocity grit channels 29–30
　by detritors 30, *31*
　by hydro-dynamic separators 41

heavy metals
　environmental quality standards for 6
　in industrial wastewaters 3, 240
　　removal by adsorbents 287–288
　limits in soils *239*
　phyto-toxic metals 238
　in sewage sludge *240*
　toxic metals 238
　toxicities 7
horizontal flow primary settlement tanks 34–36
　design criteria 36
House of Commons Environment Committee, on sludge 208
humus tanks, overflow rates for 55
hydraulic loading rates, in trickling filters 52, 54, 62
hydraulic retention times
　in activated sludge process 20, 72–73
　in anaerobic digestion reactors 261, 262
hydro-dynamic separators 41
hydrogen peroxide, oxidation by 289
hydrogen sulphide 148, 150
　exposure limits 149
　odour threshold values *148*
　removal from air 153–154
　sources 148
　typical concentrations *150*
hydrolytic bacteria 216
hydroxyl radicals, generation of 289

incineration of de-watered/dried sludge 241–245
　ash resulting from 243–244
　efficiency calculations 243
　fluidized-bed incinerators 241, *242*, 244–245
　stack emission limits *243*
　in various European countries *204*
industrial wastewaters
　see also trade effluents
　essential nutrients added 86–87

325

industrial wastewaters (*continued*)
  heavy metals in 3, 240
  quality parameters 2
  recovery of materials from 291–294
  regulations covering discharge 6–8
  treatment of 2, 273–297
    by activated sludge process 276–277
    by adsorption 286–289
    by anaerobic digestion 255
    by biological aerated filters 121–122
    by biological treatment 275–281
    by coagulants 284–285
    by deep shaft process 124
    by enhanced biological treatment 282–283, 284
    by enhanced oxidation 289–291
    by oxidation/reduction 286
    by oxygen-aerated systems 132–133
    by physico-chemical treatment 284
    by recovery processes 291–294
    by sequencing batch reactors *131*
inlet works, odour control in 151
integrated pollution control (IPC) 7, 274–275
Integrated Pollution Prevention Control (IPPC) 206
International Association of Water Quality (IAWQ) model 106–107
international biostandards, for sludge 207
iron salts
  industrial wastewaters treated with 284, 285
  phosphates removed using 169, 172, 201, 215

jet aeration
  in activated sludge process 73, 76
  in oxidation ditch 139–140
Jones activated sludge model 107

Kristianstad biogas plant (Sweden) 311–312
  characteristics *311*
  energy balance *311*
  features 311–312
Kurita process 179

lagoons
  characteristics *184*
  performance data *185*
  solids removal by 183–184
lamella plate interceptor *282*
landfill disposal of sludge 245
  in various European countries *204*
Langmuir adsorption isotherm *288*
legislation
  see also European Directives
  on potable water abstraction 4
  on trade effluents 6–8
lime
  enhanced treatment of sludge using 230
  industrial wastewaters treated with 284, 285
low pressure wet oxidation (LOPROX) process 291

Mammoth aeration plants 138, 139–140
MARPAK modular plastic media (for trickling filters) *61*
mathematical models 12–13
  for activated sludge process 105–107
  for trickling filters 64–65
mean generation time (MGT) of bacteria 19
  factors affecting 19–20
  typical values *20*
mechanical surface aerators (in activated sludge process) 73, 74–75, 83
  aeration efficiency 78
  dissolved oxygen control 83
  horizontal shaft aerators 73, 75
  in hybrid aeration systems 84
  vertical shaft aerators 73, 74, 83
media
  in biological aerated filters 118, *119*
  in nitrifying filters 163–164
  in trickling filters 44, 47–48, 60–64
membrane bioreactors (MBRs) 145–148
  characteristics *147*
  compared with activated sludge process 146–147, *147*, 314
  Kubota system *147*

sidestream compared with submerged configuration 145, *146*
Zenon system *147*
membrane filter press, de-watering of sludge by 235
membrane processes, recovery of materials using 292–294
mesophilic anaerobic digestion 216–223
metabolism 15
metal finishing effluent 3
metazoa 46, 69
methanogens 218, 253–254
kinetic constants 254
microfiltration 293–294
characteristics 292
micro-nutrients, effect on activated sludge process 313–314
microstrainers 187–188
characteristics 184
Microthrix parvicella 97, 98, 99, 103, *104*, 105
Microtox testing technique 305
Mogden formula 8
values of parameters in 9
Monod equation 20
'mudballing', in rapid gravity filters 190–191
municipal sewage
characteristics 276
quality parameters 2
municipal solid waste (MSW)
composition 309
processing of 309

nanofiltration, characteristics 292
National Research Council, trickling filter equation 64–65
nematodes 46, 69
in sewage sludge 206
Netherlands, sludge disposal options 204, 245
Nitrate Directive 3, 5, 241
nitrate pollutants 1
nitrate-nitrogen, oxidation of, by biological aerated filters 121
nitrification 46, 161–164
nitrifying filters
loading rates 162–163

media used 163–164
operational conditions 163
Nitrobacter 46, 161
growth rates 20
nitrogen content
of industrial wastewaters 276
of sludge 237, 240–241
nitrogen removal 161–168
Nitrosomonas 46, 161
growth rates 20
Nocardia spp. 97, 97, 99, 103, *104*
nutrient removal 160–180

occupational exposure limits, odour compounds 149
odour(s)
control of
in incinerators 244
in inlet works 151
in sewage treatment works 151–154
in sewers 150–151
identification of sources 152–153
legislation covering 149
measurement of 149–150
sources 148
threshold values 148
treatment of odiferous air 153–154
typical concentrations 150
oestradiol 303
oil, production from sludge 307–308
oil interceptor 281
Orbal activated sludge system 140, *140*, *141*
effluent quality from *141*, *142*
organic loading rate (OLR)
activated sludge process 70–71, 276
anaerobic digestion reactors 256, 257, 258, 261, 262, 264, 265
oxidation ditch 138
reed bed 144
sequencing batch reactor 126
trickling filter 51, *52*, 62
oxidation
of industrial wastewaters 286
of odour compounds 153–154
oxidation ditches 137–142
see also activated sludge process

327

oxidation ditches (*continued*)
  aeration devices
    Carrousel system 138, *139*
    jet aeration system 139–140
    Mammoth system 138, *139–140*
    Orbal system 140, *140, 141*
    TNO rotor 137
  effluent quality from 141–142
  mixing characteristics 138
  organic loading rates 138
  oxygen requirements 138–139
Oxy-Dep^{TM} aeration system 130,
  132–133, *132*
oxygen-aerated systems 128, 130–137
  covered tanks 133–134
  open tanks 130–133
  oxygen supply for 135–137
  sludge settlement characteristics *134*
oxygen requirements
  in activated sludge process 79–80
  in oxidation ditches 138–139
  of sewage *151*
ozone
  disinfection by 198–200
  foaming suppressed by 105
  oxidation by 283, 289
  removal of COD using 283

paper industry wastewater, treatment by
    anaerobic digestion 261, 269, 270
pathogens
  see also bacteria; nematodes; protozoa;
    viruses
  inactivation by sludge composting 246
  removal by trickling filters 55–56
percolating filters *see* trickling filters
phosphate pollutants 1
  amount from one human 2
phosphates
  removal of 168–178
    by biological treatment 173–178
    by chemical treatment 168–173,
      201, 215
  sources 168, *169*
phosphorus
  in industrial wastewaters 276
  recovery of 178–180
  in sewage sludge 237, 241

Phostrip process *174*, 175
physico-chemical treatment, industrial
    wastewaters treated by 284
'picket fence' thickener 209–210
plate-and-frame filter press, de-watering
    of sludge by 231–235
plating plant wastewater
  chemical treatment of 286, *287*
  recovery of materials from 293
poliovirus, inactivation by ozone *199, 200*
pollution 1–3
  control of 3–14
polyacrylamides, industrial wastewaters
    treated with 285
ponding in trickling filters 56–57
  causes 46, 47, 56–57
population equivalent (pe) 2–3
Portugal, sludge disposal options 204
potable water standards 4
potato wastewater, treatment by
    anaerobic digestion 261, 262, 263,
    269
powdered activated carbon (PAC),
    enhanced biological treatment
    using 282–283
precipitation, industrial wastewaters
    treated by 284–285
pre-treatment 25–43
primary settlement 34–40
  in horizontal flow tanks 34–36
  in hydro-dynamic separators 41
  multiple weirs in 39, *40*
  in radial flow tanks 36–38
  weir loadings for 39
primary treatment, with long sea outfall,
    effect on bacteriological water
    quality *42*
protozoa 21–23, 46, 69, 206
pyrolysis of sludge 308–309

radial flow primary settlement tanks
    36–38
  design criteria 36
  scum troughs in 38
  stilling chamber in 37–38
rapid gravity filters 188–191
  characteristics *184*
  compared with deep bed filters *192*

readily biodegradable chemical oxygen
  demand (RBCOD) 167
  values quoted 167
re-cycling of sludge 306
reduction, industrial wastewater treated
  by 286
reed beds 142–145, 194–195
  compared with grass plots 186
  effluent quality after 144, *195*
  horizontal-flow systems 143–144
  organic loading rate for 144
  roles 142
  with rotating biological contactors
    145, *195*
  for secondary treatment 143–144
  species used 142
  for storm sewage 33, 194
  for tertiary treatment 194–195
  vertical-flow systems 144, *194*
Renovexx filter 294
reverse osmosis
  characteristics *292*
  recovery of materials using 293
rheological characteristics, sewage sludge
  213–215
Rhodococcus rhodochrous, reagents
  affecting foaming ability 313, *314*
River Ecosystem (RE) classes 10
  parameters defining *11*, *13*
river quality objectives (RQOs) 3
rotating biological contactors (RBCs)
  113, 114–115, *116*
  compared with submerged biological
    contactors 115, *116*
  with reed beds 145, *195*
rotifers 46, 69

Safe Sludge Matrix 208, 237
salmonella bacteria
  inactivation by ozone 199
  limit for bathing water 196
  limit for potable water 4
  in sewage sludge 206
screenings
  amount per day 25
  composition 26
  cost of disposal 25
  maceration of 27

washing/pressing system for *27*, *28*
screens 25–29
  design of 29
  initial removal of solids on 26
scum 38
  removal of 38
scum trough(s) in primary settlement
  tanks 38
sea outfalls 41–43
  design criteria 42
  post-commissioning survey advisable
    43
sensitive waters
  criteria for 160
  in UK 161
septicity of sewage feed, effect on
  activated sludge settleability 85
sequencing batch reactors (SBRs)
  126–128
  aeration of 127
  characteristics *130*
  decanters used 127, *128*, *129*
  design criteria 126–127
  in enhanced biological treatment 283
  numbers in Bavaria *130*
settlement tanks
  in activated sludge process 87–90
  primary settlement tanks 34–40
  for trickling filters 55
sewage biosolids 305–312
sewage sludge 204
  see also sludge...
  amount produced 204
  anaerobic digestion of 216–228
  calorific values *242*
  carbon:nitrogen ratio 246
  composition 204, *237*, 240–241, *240*
  de-watering of 230–237
  disposal of 204, 237–247
  enhanced treatment of 228–230
  filterability 233
  gasification/pyrolysis of 308–309
  heavy metals in 238–240
  nitrogen content *237*, 240–241
  phosphorus content *237*, 241
  rheological characteristics 213–215
  thermal drying of 247–248
  thickening of 209–213

sewage treatment works
    odour control in 151–154
    processes in 13, 14
sewers
    odour control in 150–151
    UK system 3
sheep dips, pollution by 297
silage 295
silage liquors 294, 295–296
SIMPOL model 12–13
slime layer (in bacteria) 16
sludge de-watering 230–237
    by belt press 236–237
        effect of cationic polymer 235
    by filter press 231–235
sludge disposal
    on agricultural land 237–241
    by composting 245–247
    by incineration 241–245
sludge growth index, effect of loading rate 71
sludge handling and disposal 204–248
    factors affecting 204–209
    options listed 205
    routes in various European countries 204
sludge loading rate 70
    see also organic loading rate
sludge rheology 213–215
sludge settleability 85
    factors affecting 85–87
sludge thickening 209–213
    by belt thickeners 211
    by centrifugation 210, 211
    by dissolved air flotation 212–213
    by gravity thickening 209–210
Sludge (Use in Agriculture) Regulations (1989) 206, 237
sludge volume index (SVI) 85
slurried animal wastes
    BOD data 294
    treatment of 296–297
solids removal
    by primary settlement 34–40
    by screens 25–29
    by tertiary treatment 183–195
    grit removal 29–31

sonication, anaerobic digestion enhanced by 226–227
Sphaerotilus natans 17, 95, 99
sporozoa 23
spring sloughing (in trickling filters) 46, 58
staining of bacteria 21, 95
statutory nuisance, legal definition 149
stirred specific volume index (SSVI) 85
    and design of activated sludge final settlement tanks 89–90
    and length of filamentous microbes 99, *100*
STOAT software 12, 64
Stoke's Law 281
storm sewage
    reed beds used for 33, 194
    treatment of 33
storm tanks 33
    design criteria for 33–34
storm water separation 32–34, 41
struvite (magnesium ammonium phosphate)
    formation during phosphate removal 177, 178
    uses 178
sub-critical wet oxidation 307
submerged aerated filters (SAFs) 122
submerged biological contactors (SBCs) 113, 115–116, *117*
    compared with rotating biological contactors 115, *116*
submerged filters 117–122
sulphate-reducing bacteria (SRBs) 254–255
super-critical wet oxidation 307
Surface Water Abstraction Directive 4, 5
suspended solids (SS)
    amount from one human 1
    in industrial wastewaters 2
    in municipal sewage 2
    removal of
        by anaerobic reactors 263, 268
        by biological contactors *116*
        by grass plots *185*
        by membrane bioreactors *147*
        by primary settlement tanks 34

by rapid gravity filters 190, *191*
by reed beds 144, *195*
by tertiary treatment lagoons *184, 185*
swing adsorption, oxygen-rich air generated by 135–136
Syngas, calorific value 309

tertiary treatment 183–202
  disinfection 195–202
  solids removal 183–195
    characteristics of various processes *184*
    by grass plots 184–186
    by lagoons 183–184
    by microstrainers 187–188
    by rapid gravity filters 188–191
    by reed beds 194–195
    by upflow clarifiers 186, *187*
    by upflow sand filters 193–194
Tetra filter 192
  de-nitrification by 168
thermal drying of sludge 247–248
thermal treatments, for sludge processing 306–307
thermophilic aerobic digestion (TAD) process 228, 230
  handling characteristics of sludges 229–230
thermophilic anaerobic digestion 227
  in two-stage systems 224–225
thickening of sludge 209–213
  see also sludge thickening
Thiothrix spp. 97, 99
TNO rotor (in oxidation ditch) 137
toxicity testing techniques 305
toxicological effects 304–305
trade effluents
  see also industrial wastewaters
  characteristics 276
  charging policy for 8–10
  compounds restricted in 8
  consents for 6–7, 281
  control of discharge 6–8
  costs of treatment 9
  effect on nitrification 162
treatment processes *13, 14*
trickling filters 44–65

algae and fungi in 46
alternating double filtration mode 52–55, *54*
biofilm on packing 45–47
characteristics of media used 48, 62
characteristics of process 44
circular 49–50
compared with activated sludge process 44
design criteria 51
diffusion processes in 47
distributors for 49–50
double-filtration mode 52, *53*
filtration with re-circulation 51, *53*
final settlement tanks for 55
flies on 57–58
high-rate filtration 60–63
hydraulic loading rates 52, 54, 62
low-rate filtration 63–64
media used 44, 47–48, 60–64
models for performance-prediction 64–65
modes of operation 50–55
moss on 57
operational problems 56–60
organic loading rates 51, 52, 62
overflow rates for humus tanks 55
packing materials 44, 47–48
performance 55–56
phosphate removal from effluent 170
plastic media used 60–64
  in high-rate filtration 60–63
  in low-rate filtration 63–64
ponding on 46, 47, 56–57
rectangular 50
single-pass filtration mode 50–51
spring sloughing in 46, 58
temperature effects 58–60
tower configuration 62
ventilation of 48, 62–63
triglycerides 278
trihalomethanes (THMs) 198
Trojan ultraviolet system 201
tube viscometer (for sludge) 213–214
Type 021N microbe 97, 98, 99

UK, sludge disposal options 204, 237, 243, 244, 245

ultrafiltration, characteristics 292
ultrasound, anaerobic digestion enhanced by 226–227
ultraviolet (UV) radiation, disinfection by 200–202
UNOX activated sludge process 133–134
  performance 135
upflow anaerobic filter 257, 258
  performance 258
upflow anaerobic sludge blanket (UASB) reactors 258–261
  granular sludge formed in 259–260
  hybrid reactor 261
  performance data 261
upflow clarifiers 186, 187
  types 186
upflow sand filters 193–194
urban pollution management (UPM) 10–14
Urban Wastewater Treatment (UWWT) Directive 3, 4–5
  effects on sludge disposal 204–205
  on eutrophic conditions 160
  standards required by 5, 161, 176
  treatment required by 5, 41

US Environment Agency, regulations on sludge 207

ventilation, of trickling filters 48, 62–63
vermiculture 306
  see also composting
VerTech wet oxidation plant 307, 308
viruses 23
  removal by membrane bioreactors 146
  in sewage sludge 206
viscometers (for sludge rheology) 213–214
Vitox^{TM} aeration system 130, *131*, 132
vortex separators 33

Water Industry Act (1991) 6, 7–8
Water Research Centre, control sequence for bilking sludges 99, *100*, 101
water sprays, to suppress foaming 105
wet air oxidation (WAO) 290–291, 307
wet chemical scrubbers, for odour control 153
Whole Effluent Toxicity (WET) 304